Archaeological Chemistry

Curt W. Beck, *Editor*

A symposium sponsored by
the Division of the History
of Chemistry at the
165th Meeting of the
American Chemical Society,
Dallas, Tex., April 9-10, 1973.

ADVANCES IN CHEMISTRY SERIES **138**

AMERICAN CHEMICAL SOCIETY
WASHINGTON, D. C. 1974

Library of Congress CIP Data

Archaeological chemistry
(Advances in chemistry series, 138)

Symposium on Archaeological Chemistry, 5th, Dallas, 1973.
Includes bibliographical references and index.

1. Archaeology—Methodology—Congresses. 2. Archaeological chemistry—Congresses.
I. Beck, Curt W., 1927- ed. II. American Chemical Society. Division of History of Chemistry. III. Title. IV. Series.

QD1.A355 no. 138 [CC75] 540'.8s [913'.031'0285]
74-22372
ISBN 0-8412-0211-7 ADCSAJ 138 1-254 (1974)

PRINTED IN THE UNITED STATES OF AMERICA

Advances in Chemistry Series

Robert F. Gould, *Editor*

FOREWORD

ADVANCES IN CHEMISTRY SERIES was founded in 1949 by the American Chemical Society as an outlet for symposia and collections of data in special areas of topical interest that could not be accommodated in the Society's journals. It provides a medium for symposia that would otherwise be fragmented, their papers distributed among several journals or not published at all. Papers are refereed critically according to ACS editorial standards and receive the careful attention and processing characteristic of ACS publications. Papers published in ADVANCES IN CHEMISTRY SERIES are original contributions not published elsewhere in whole or major part and include reports of research as well as reviews since symposia may embrace both types of presentation.

CONTENTS

Preface vii

Introduction ix

1. **Isotopic Analysis of Laurion Lead Ores** 1
   I. L. Barnes, W. R. Shields, T. J. Murphy, and R. H. Brill

2. **Analysis of Sasanian Silver Objects: A Comparison of Techniques** 11
   D. F. Gibbons, K. C. Ruhl, and L. S. Staikoff

3. **Major and Trace Elements in Sasanian Silver** 22
   Pieter Meyers, Lambertus Van Zelst, and Edward V. Sayre

4. **Archaeological Ceramic Study Possibilities with a Thermal Gradient Furnace** 34
   Frederick R. Matson

5. **Biblical Studies through Activation Analysis of Ancient Pottery** . . . 48
   D. Brooks, A. M. Bieber, Jr., G. Harbottle, and E. V. Sayre

6. **Correlation between Terra Cotta Figurines and Pottery from the Valley of Mexico and Source Clays by Activation Analysis** 81
   R. Abascal-M, G. Harbottle, and E. V. Sayre

7. **Neutron Activation Analytical Survey of Some Intact Medieval Glass Panels and Related Specimens** 100
   Jacqueline S. Olin and Edward V. Sayre

8. **Neutron Activation Analysis of Gold Impurity Levels in Silver Coins and Art Objects** 124
   Adon A. Gordus and Jeanne P. Gordus

9. **Comparative Analysis of Archaeological Bronzes** 148
   W. T. Chase

10. **Mössbauer Effect Spectroscopy without Sampling: Application to Art and Archaeology** 186
    Bernard Keisch

11. **Preliminary Investigations on Chinese Ink in Far Eastern Paintings** 207
    John Winter

12. **Nuclear Magnetic Resonance Spectrometry in Archaeology** 226
    Curt W. Beck, Constance A. Fellows, and Edith MacKennan

13. **Chemical Investigations of Ancient Near Eastern Archaeological Ivory Artifacts** 236
    N. S. Baer and N. Indictor

Index 247

PREFACE

In his introduction to the fourth symposium on archaeological chemistry ("Science and Archaeology," M.I.T. Press, 1971), Robert H. Brill states that archaeological chemistry is emerging from the phase of a service science to the field archaeologist and the museum curator into a discipline of its own. The papers collected here from the fifth symposium support this view.

By its very nature, archaeological chemistry has had its own distinct parameters all along; the unique quality and the value of the objects analyzed have always been a goad towards the development of micromethods or, ideally, of completely non-destructive methods. As the number of analyses of artifacts increases and the composition of large groups of artifacts reveals patterns which are becoming more and more useful for determining the time, place or method of manufacture, other specific considerations emerge: accuracy and precision in the context of their archaeological significance. The choice of analytical method is often not determined by the greatest accuracy attainable but by the optimal means of obtaining, rapidly and economically, the accuracy needed to make archaeologically meaningful distinctions.

Methodological studies inevitably lead to some revision and even rejection of past work. Archaeologists who were too readily persuaded to accept scientific results as absolute facts may be a little dismayed to find that some of these "facts" are now being questioned. They must remember that no discipline has a corner on infallability and that knowledge, in the sciences as well as in the humanities, advances slowly by a kind of iterative process in which error is gradually reduced to reach an approximation of truth. Uninhibited self-criticism is the surest indication that archaeometry is coming of age.

Along with these welcome signs of maturity there is continuing evidence of vigorous growth. Each year, new physical and chemical analytical methods are applied to old archaeological problems, and new types of archaeological materials are investigated by established analytical methods. This volume contains examples of both of these directions of expansion.

A symposium on any rapidly growing field can only be a snapshot, showing one moment of development and not even all of that. Physical methods of dating and archaeological prospecting are only two important branches of archaeometry which fall outside the focus of this volume. Since artifact analysis is the particular province of chemists, it is especially appropriate that this book should appear in the ADVANCES IN CHEMISTRY SERIES.

CURT W. BECK

Vassar College
February 1974

INTRODUCTION

The Division of the History of Chemistry is honored to sponsor this Fifth Symposium on Archaeological Chemistry. Like its predecessors, it is the creation of its organizers, its chairman, and its participants, and we are grateful to those who have made this event possible.

Specifically, I thank Robert Brill, of the Corning Museum of Glass, who organized the symposium, and who would have been its Chairman had it not been for the 1972 flood resulting from Hurricane Agnes, which ravaged the museum and made it necessary for him to devote all his time to repairing the damage. I also want particularly to thank Curt W. Beck, Professor of Chemistry at Vassar College, who agreed to take over the chairmanship. All who attended the symposium or who read this printed volume of its proceedings will appreciate the work of the authors, and the Division adds its thanks.

Ever since the end of the 18th Century, when Klaproth analyzed some ancient glasses, chemists in increasing numbers have been fascinated by the things that chemical analysis can tell about ancient history, ancient ways of life, including the technical processes and the chemical substances used by the ancients, and patterns of trade in the ancient world.

When I was a graduate student at Princeton, E. R. Caley was a young instructor there. I have never forgotten my delight at his description of how the decline in the standards of honesty of the Roman Empire was paralleled by and reflected in the steady rise in the lead content of its bronze coins, or my absorption in his tales of tracking down forgeries by chemical analysis. As analytical methods have improved, archaeological chemists have been able to furnish more and more information about more and more specimens. X-ray diffraction analysis, and the new, nondestructive methods of analysis are cases in point. How far we have come will be apparent to readers of the papers collected here and even more to those who have read the papers from the previous four symposia, the subject of which might well be called the "chemistry of history."

PETER OESPER

St. Lawrence University
Chairman, Division of the History of Chemistry

1

Isotopic Analysis of Laurion Lead Ores

I. L. BARNES, W. R. SHIELDS, and T. J. MURPHY

Analytical Chemistry Division, Institute for Materials Research, National Bureau of Standards, Washington, D. C. 20234

R. H. BRILL

The Corning Museum of Glass, Corning, N. Y. 14830

The lead isotopic ratios of a carefully selected suite of ore samples from the Laurion region have been determined by a precise mass spectrometric procedure. The ores were taken from various levels in mines, some known to have been worked in ancient times. All are nearly indistinguishable isotopically within the precision of the method (±0.05%), and they closely match leads from archaeological objects found in Greece. A comparison with isotopic data for ores from other mining regions in the ancient world has been made. The uniformity of the Laurion ores facilitates the interpretation of lead isotope data for archaeological objects from Greece.

Chemical analysis of archaeological objects has been done for many years and, where practiced with care, has resulted in information of great value. This has been amply illustrated at several symposia over the past decade. While there probably will never be an entirely satisfactory substitute for complete chemical analysis to characterize early objects, this method alone does not ordinarily tell us all we would like to know about date, origin, authenticity, or method of manufacture. Other approaches are equally valuable, and there is a growing recognition of the complementarity of chemical analysis, physical testing, microscopic examination, and other specialized techniques appropriate for particular materials. A striking example of this approach is the investigation of the famous bronze horse in the collection of the Metropolitan Museum of Art. It has been subjected to a dozen or more different types of examinations, as demonstrated in a recent exhibition at the Museum.

Lead Isotopic Analysis

Lead isotopic analysis (*1, 2, 3, 4, 5*) is a relatively new member of this expanding group of complementary methods. Like the others, it has its own peculiarities, advantages, and disadvantages which have been described. Lead isotope studies help to identify possible geographical origins of the ores from which ancient leads were smelted. Even where specific mines cannot be identified, the objects can still be classified as to which could or could not have had a common origin. The method requires only small sample sizes, and the results are independent of the histories of the objects studied.

The usefulness of lead isotope studies is the fortunate result of a threefold coincidence. Historically, lead was one of the first metals mined—its first use dating back at least into the fourth millennium. Thus, it occurs in artifacts associated with many historical places and periods. Technologically, it was easily won from its major ore, galena, and was quickly found to be useful in many forms: alloys, pigments, cosmetics, medicines, glasses, glazes, and others. Since it was also produced in large amounts as a by-product of silver extraction, it was abundant and relatively inexpensive. Geologically, and most importantly, lead is one of the few chemical elements which varies markedly in its istotopic makeup according to its geological occurrence. Putting these three factors together, one has the basis of a very useful archaeometric tool.

The element lead is composed of four istotopes—^{208}Pb, ^{207}Pb, ^{206}Pb, and ^{204}Pb. The first three are the stable end products of the radioactive decay of uranium and thorium.

$$^{238}\text{U} \rightarrow {}^{206}\text{Pb} \qquad T_{1/2} = 4.4683 \times 10^9 \text{ years}$$
$$^{235}\text{U} \rightarrow {}^{207}\text{Pb} \qquad T_{1/2} = 0.70381 \times 10^9 \text{ years}$$
$$^{232}\text{Th} \rightarrow {}^{208}\text{Pb} \qquad T_{1/2} = 14.01 \times 10^9 \text{ years}$$

Because of the significant differences in the half-lives of the above reactions, leads in different geological environments acquire different (although not necessarily unique) isotopic compositions, which are continuously changing. In addition, different and varying amounts of lead, uranium, and thorium in different geological environments throughout geological time resulted in even larger changes. However, if during the process of ore formation the lead separates from its parent uranium and thorium (as almost always happens), its isotopic composition becomes "frozen" and ceases to change, leading to a fixed composition within a given deposit. When the various isotopic ratios can be measured with sufficient precision, different ore deposits can be distinguished and matched with objects made from or containing leads from those sources.

The value of this technique has been firmly established even though the earliest work was hampered by instrumentation and methods which could not yield highly precise data and were tedious and time-consuming experimental procedures.

Work by Catanzaro *et al.* in 1968 (*6*) led to a new analytical procedure permitting the measurement of isotopic ratios to about ±0.05% (95% L.E.); this resulted in the availability of three standard reference materials, so that results could be placed on an absolute basis. This procedure, still the most precise and accurate one available, requires about 1 mg of lead for an analysis. A second procedure (*7*) has been developed which utilizes silica gel as an ionization enhancer. This method permits the measurement of isotopic ratios to about ±0.1% (95% L.E.), but it requires only 0.1 μg of lead per analysis. In addition, the instrumentation and data handling have been vastly improved so that many samples can be studied quickly and conveniently.

Our Lead Studies

These developments have encouraged us to undertake a long term study of objects, both of known and unknown origins, and of lead ores from mining areas worked in ancient times. Our research is now ranging into all sorts or archaeological and historical materials—*e.g.*, glasses, glazes, bronze coins, white-lead pigments, traces of lead in silver and gold, Chinese bronzes, ancient Egyptian kohl, and the caming in medieval stained glass windows. This has all been made possible by the development of high precision instrumentation.

We have analyzed lead ores (mostly galenas) from many countries, but our ore data are still fewer than desired. The problem is to obtain well-documented ore samples from areas known or suspected to have been exploited in ancient times. The present study is an effort to correct this, along with another related facet of the ore problem—*i.e.*, just how much isotopic variation exists within given mining regions. While our choice of a region may not be representative, it does provide data on one of the most important lead mining regions of the ancient world, the Laurion mines in Greece.

We have been fortunate in acquiring a suite of carefully selected ore specimens from three mining sites within the region customarily described as "Laurion." These samples, including sulfide and oxide ores, were collected to represent the general area and various contact levels within the mines sampled. The mines, located about 20 km south of Athens, are designated by the present operators as the Plaka, Esperance, and Kamarisa mines and are separated by a few kilometers. Of the 17 ore specimens analyzed, about one-half are from occurrences immediately adjacent to ancient workings.

These samples are particularly valuable for this study since the area is undoubtedly the source of much of the lead used in early times. It is not known exactly when mining first began there, but it was probably before 1000 B.C. Marinos and Petrascheck (*8*) have described the geology of the region, the nature of the ore deposits, and the history of mining at Laurion. According to them, the most important period of exploitation was from 483 B.C. onward, when the deposits at present day Kamarisa were discovered. They state that there are today more than 1000 ancient mine shafts and galleries, some of which are more

Table I. Lead

NBS No.	*Brill No.*	*No. Sample*	*Mine*	*Level*	*208/206*
383	850	1 sulfide A[a]	Plàka	145	2.0600
384	851	1 sulfide B	Plaka	145	2.0604
385	852	2 sulfide A	Plaka 33	110	2.0605
386	853	2 sulfide B	Plaka 33	110	2.0602
388	854	3 sulfide A	Esperance	96	2.0611
387	855	3 sulfide B	Plaka 33	110	2.0637
389	856	3 oxide A[b]	Esperance	103	2.0600
390	857	3 oxide B	Esperance	103	2.0606
391	858	4 sulfide A	Plaka Filon 80	85	2.0596
392	859	4 sulfide B	Plaka Filon 80	85	2.0591
393	860	4 sulfide C	Plaka Filon 80	85	2.0598
394	861	4 oxide A	Plaka Filon 80	135	2.0590
395	862	4 oxide B	Plaka Filon Sklives	80	2.0614
396	863	5 sulfide A	Kamarisa	135	2.0593
397	864	5 sulfide B	Kamarisa	135	2.0594
398	865	5 sulfide C	Kamarisa	135	2.0580
399	866	6 sulfide A	Kamarisa	104	2.0579

[a] Major lead mineral in the ore samples identified as "sulfides" is galena (PbS).
[b] Ore samples identified as "oxides" are highly oxidized or gossan ores in which the major lead mineral is also galena.

than 100 m deep and that mining was carefully planned since the ancient miners quite early understood that the ore was to be found in the marbles, especially at their contact planes with the schists. Some other key references on the history of the area are those of Ardaillon (*9*), Davies (*10*), Boulakia (*11*), and Hopper (*12*).

Experimental

The ore samples were carefully crushed in a mortar and weighed into Teflon beakers. They were dissolved in 10 ml of 1:1 (vol) perchloric acid; this acid solution was diluted with water to 100 ml, and the lead was removed as PbO_2 by electroplating at 2.0 amps for 16 hours. The lead was then removed from the anode with 1 ml of a solution of nitric acid (2%) and hydrogen peroxide (0.3%). After heating to destroy the peroxide, the lead was taken up in sufficient 2% nitric acid to make a solution equivalent to 25 mg of lead per gram of solution, which was reserved for mass spectrometric analysis.

The analytical procedure was the mass spectrometric triple filament procedure of Catanzaro (*13*). A drop of the sample was placed on each of two side filaments of rhenium, lead hydroxide was precipitated on the filaments by adding a drop of ammonium hydroxide, and the precipitate

Isotopic Ratio Data

		Atom Percent			
207/206	*204/206*	*208*	*207*	*206*	*204*
0.83127	0.053005	52.228	21.075	25.353	1.3439
0.83143	0.053104	52.229	21.076	25.349	1.3461
0.83175	0.053103	52.226	21.082	25.346	1.3460
0.83196	0.053134	52.219	21.087	25.347	1.3468
0.83210	0.053099	52.229	21.086	25.340	1.3455
0.83201	0.053100	52.261	21.070	25.324	1.3447
0.83149	0.053061	52.224	21.080	25.351	1.3452
0.83128	0.053047	52.234	21.072	25.349	1.3447
0.83120	0.053030	52.223	21.076	25.356	1.3446
0.83044	0.052984	52.228	21.064	25.365	1.3439
0.83098	0.052975	52.229	21.071	25.357	1.3423
0.83131	0.053049	52.214	21.081	25.359	1.3453
0.83204	0.053054	52.234	21.083	25.339	1.3443
0.83218	0.053129	52.205	21.097	25.351	1.3469
0.83165	0.053073	52.214	21.086	25.354	1.3456
0.83170	0.05298	52.198	21.095	25.364	1.3438
0.83083	0.053011	52.208	21.078	25.370	1.3449

was dried and mounted in the instrument along with a platinum ionizing filament. At least 10 measurements of each of the three ratios were made in a 12-inch radius mass spectrometer.

Although with samples as large as these, blanks are so small as to be negligible, precautions were taken to prevent unnecessary contamination. All processing was carried out in a Class 100 clean laboratory, and subsequent handling was done in small Class 100 clean air hoods. All equipment was specially cleaned, and all reagents were purified by a sub-boiling distillation technique known to reduce lead contamination to very low levels (*14*). The normal laboratory lead blank is 1-2 ng per analysis.

Results

Data obtained for the 17 samples are shown in Table I, along with the computed compositions (atom percent) of each sample. The data are also shown in Figure 1 as a plot of the 208/206 *vs.* 207/206 ratios. Note the remarkable uniformity shown by all of these samples. The maximum variation of the 208/206 ratios is less than 0.3% while that of the 207/206 ratios is less than 0.2%. With our expected precision of ±0.05%, this indicates that while the samples show small but statistically significant differences, they are very similar and must have been derived from the same or very similar source rocks in a relatively short time. The real differences observed are most likely related to slight

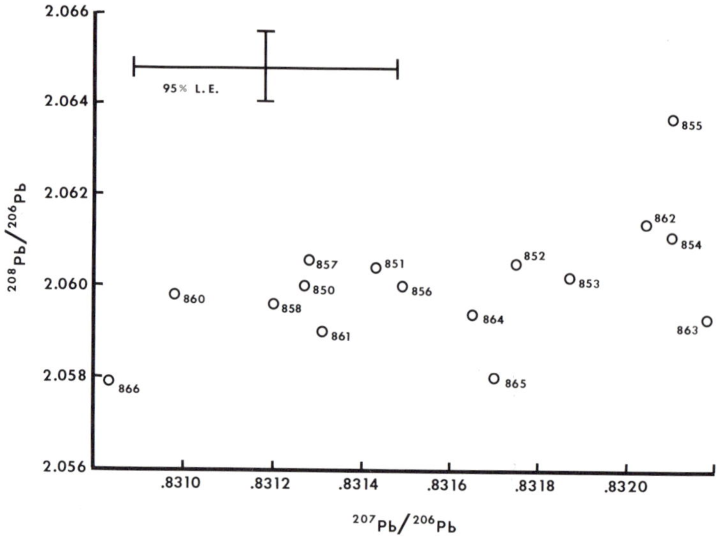

Figure 1. $^{208}Pb/^{206}Pb$ vs. *$^{207}Pb/^{206}Pb$ for some Laurion lead ores. Numbers for each point are given in Table I. Error bars represent the 95% limit of error for each analysis.*

variations in the source of the ore fluids, contamination of the ore fluids by country rock, and to slight variations in time of the extraction of the ore fluids. Figure 2 is a plot of 208/206 ratio *vs.* the depth of the deposit from which the sample was obtained and for each of the three separate areas. No real correlations are obvious, nor is there any marked difference between the sulfide and oxide ores.

The narrow range of lead isotopic abundances in the Laurion area as shown in this study and the fact that major amounts of lead from this area were mined and used in ancient times indicate that these ratios should serve as a reasonable tracer. The data also justify the assumption previously made, but until now unproved, that the lead output of the Laurion mines was essentially of one uniform and recognizable type.

We have analyzed about 20 ancient objects, which we know (with near certainty) were made of Laurion lead. For example, one of our reports (*15*) gives the analytical results of a series of seven early coins minted in Athens. We had concluded that these probably contained locally obtained lead, and we are now confident that this unquestionably is true. It also appears that Laurion lead was exported since some 2nd century B.C. coins minted in Alexandria and Syria contain this same type of lead. Other objects we have analyzed indicate that Laurion lead was used in the 6th century B.C. Material from the handles of bronze vessels dated from the 6th, 5th, and 4th centuries B.C. all show isotopic ratios in the narrow range characteristic of this lead (*1*). One in particular, which had been attributed to either Sparta or southern Italy shows lead ratios nearly identical to the average for Laurion leads reported here and much different from any ratios which we have found in Italian leads. A few earlier specimens, of Mycenaean date, are also of the Laurion type. However, lead deposits in the nearby Greek islands, for example on Kea, might have ratios very close to those of the Laurion ores. Until we analyze samples from such sites, we cannot assume either

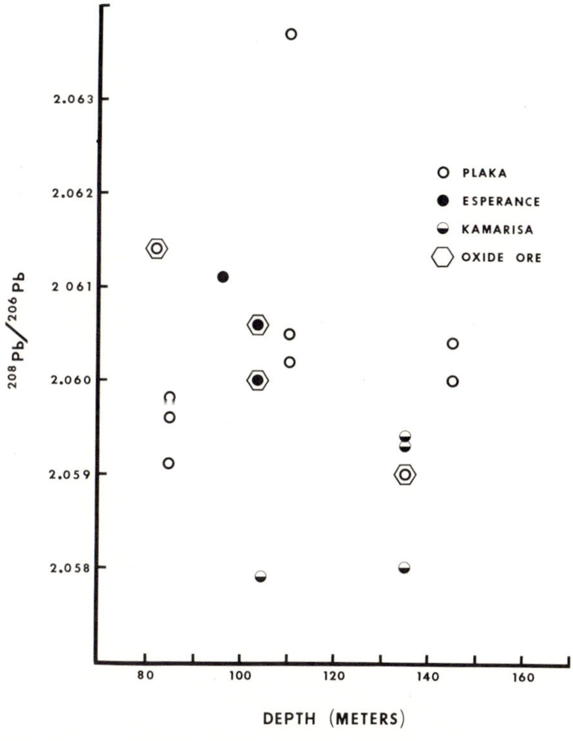

Figure 2. $^{208}Pb/^{206}Pb$ ratio vs. *the depth in meters for the Laurion lead ore samples. Points surrounded with a hexagon represent oxide ores. Other samples are sulfide ores.*

that they differ significantly from those discussed here or that they are identical to them.

Future Work

Another area in need of investigation is northern Greece, but we have been unable to obtain suitable ore specimens from that region. Lead sheathing from the roof of the monastery at Iviron on Mt. Athos

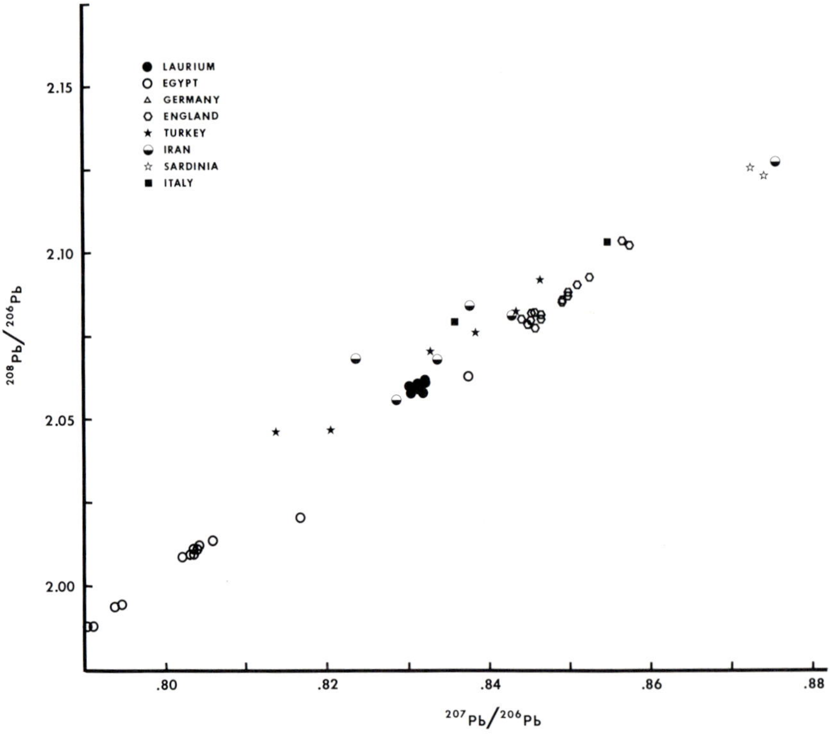

Figure 3. Relationship between the ores of Laurion and those of other areas are shown in a plot of $^{208}Pb/^{206}Pb$ vs. $^{207}Pb/^{206}Pb$ *ratios. Each point represents the approximate size of the 95% limit of error for the individual analysis.*

has an isotopic composition unlike the Laurion lead; this fact reminds us that just because a lead is found within the present-day political boundaries of Greece, it need not be of the Laurion type. This lead looks much more like a group of leads we associate with the general region of Istanbul. The sheer massiveness of the monastery sheathing strongly suggests that it probably is a product of the nearest mines, either in northern Greece or the mines supplying the environs of Istanbul.

Although it is more interesting to analyze lead from actual objects of archaeological importance rather than performing repeated analyses of galena, in any study which attempts to define the origin of raw materials much effort must be spent in determining the location and the expected variation of the material from the source area. To date we have been able to do this for only a few areas—*e.g.*, England, to a lesser extent, Turkey, and now quite thoroughly for Laurion. Our knowledge of the isotopic composition of lead ores from Italy, northern Greece, Spain, Mesopotamia, India, and China is still inadequate—a situation we hope to correct. Laurion leads are compared in Figure 3 with those found in some of these other areas.

Previously we have tended to treat our findings on objects empirically, noting which objects appeared to have common origins and which did not. As our catalog of ore data grows, we are more confident in attributing objects to specific mining regions. We hope that geologists or archaeologists who have access to ore specimens will contribute to this catalog.

Acknowledgment

The authors are indebted to the many persons who have provided samples for analysis. We particularly thank J. M. Lambert, Director Compagnie Francaise Des Mines Du Laurium and his colleagues, Mr. Papathanassiou, Mr. Miret, and Mr. Sychowicz who provided the excellent ore samples and to the persons mentioned in our previous catalogs who provided the ancient specimens. We also thank G. N. Hanson who performed many of the analyses while a guest worker at NBS.

Literature Cited

1. Brill, R. H., Wampler, J. M., "Isotope Studies of Ancient Lead," *Amer. J. Archaeol.* (1967) **LXXXI**, 63-77.
2. Brill, R. H., Wampler, J. M., "Isotope Ratios in Archaeological Objects of Lead," in "Application of Science in Examination of Works of Art," W. J. Young, Ed., pp. 155-66, Boston Arts Museum, Boston, 1967.
3. Brill, R. H., "Lead Isotopes in Ancient Glass," *Ann. Congr. Journees Int. Verre, 4th,* (International Association for the History of Glass), *Liège, 1969*, 255-51.
4. Brill, R. H., "Lead and Oxygen Isotopes in Ancient Objects," in "The Impact of the Natural Sciences on Archaeology," T. E. Allibone *et al.*, Eds., pp. 143-64, London, 1970; same paper in *Phil. Trans. Roy. Soc. Lond. A.* (1970) **269**, 143-64.
5. Brill, R. H., Shields, W. R., Wampler, J. M., "New Directions in Lead Isotope Research," in "Application of Science in Examination of Works of Art," W. J. Young, Ed., Boston Arts Museum, Boston, 1971.
6. Catanzaro, E. J., Murphy, T. J., Shields, W. R., Garner, E. L., *J. Res. Nat. Bur. Stand., Sect. A* (1968) **72A** (3), 261-267.

7. Barnes, I. L., Murphy, T. J., Gramlich, J. W., Shields, W. R., "Lead Separation by Anodic Deposition and Isotope Ratio Mass Spectrometry of Microgram and Smaller Samples," *Anal. Chem.* (1973) **45,** 1881-1884.
8. Marinos, G. P., Petrasheck, W. E., *Geol. Geophys. Res.* (Athens) (1956) **4,** 223-36.
9. Ardaillon, E., "Les Mines du Laurion dans l'Antiquité," Paris, 1897.
10. Davies, O., "Roman Mines in Europe," Oxford, 1935.
11. Boulakia, J., "Lead in the Roman World," *Amer. J. Archaeol.* (1972) **76** (2), 139-44.
12. Hopper, R. J., "The Laurion Mines: A Reconsideration," *Ann. Brit. School Athens* (1968) **63,** 293-326.
13. Catanzaro, E. J., *Earth Planet. Sci. Lett.* (1967) **3,** 343-46.
14. Kuehner, E. C., Alvarez, R., Paulsen, P. J., Murphy, T. J., *Anal. Chem.* (1972) **44,** 2050-56.
15. Brill, R. H., Shields, W. R., "Lead Isotopes in Ancient Coins," in "Methods of Chemical and Metallurgical Investigations of Ancient Coinage," E. T. Hall, D. M. Metcalf, Eds., pp. 297-304, Royal Numismatic Society, London, 1972.

RECEIVED July 9, 1973.

2

Analysis of Sasanian Silver Objects: A Comparison of Techniques

D. F. GIBBONS, K. C. RUHL, and L. S. STAIKOFF

The Center for the Study of Materials, Case Western Reserve University, Cleveland, Ohio 44106

We obtained data to demonstrate whether the metallurgical structure affected the accuracy with which copper, gold, and lead concentrations could be determined on small samples (< 1 gram) from ancient silver specimens. The data were taken from objects mainly assigned to the Sasanian period. Analytical methods included x-ray fluorescence, electron microprobe, and thermal neutron activation analyses. The data demonstrated that the accuracy is directly related to the uniformity of the metallographic structure. Surface sampling techniques tend to underestimate the copper concentration, and care must be taken in interpreting gold analyses from surface rubbing techniques to ensure that the object had not previously been gilded. Emission spectroscopic analyses tend to overestimate the lead concentration.

Elemental analysis, especially utilizing nondestructive techniques, is a natural method to consider when studying the technology used in ancient works of art. The chemical composition of a metal object can be directly related to two essential metallurgical processes—refining and fabrication. For silver-based alloys the concentration of gold and lead in the alloy is related to the sophistication of the silver refining technology as well as the composition of the ore from which it was refined. The concentration of copper alloyed with the silver is inversely related to the malleability or ease with which the alloy can be hammered into a desired shape and directly related to the fluidity or ease with which fine detail can be reproduced in casting. This paper is concerned with the major and minor alloying elements—copper, gold, and lead—in Sasanian and related silver objects. We are arbitrarily defining elements in concentrations less than 0.1 wt % as trace elements. Trace elements

Table I. Classification of Samples

Object	*Class*	*Source*
Oval Bowl	I	Private collection of D. Shepherd.
King Torso	I	Insert from unknown hunting plate. Private collection of D. Shepherd.
Goblet from Vicarello, Italy	I	Ausustan period. CMA[a] 66.371.
Horse Rhyton	II	CMA[a] 64.41.
Anahita Rhyton	II	CMA[a] 62.294.
Silver Wire	II	From hinge of Ewer. Private collection.
Plate with Bacchanalian Scene	IIA	Freer[b] 64.10.
Bull Bowl	III	Cast bowl, Achaemenid period. Private collection.
Silver Nail	III	Cast. Private collection.
Horse Rhyton	II	Private collection.
Female Head	IIIA	Two samples from different parts of object. Private collection.
King's Head	IIIA	Freer[b] 66.23
Queen's Head	IIIA	Freer[b] 66.24.
Hormizd Hunting Plate	IIA	CMA[a] 62.150.
Stroganoff Hunting Plate	IIA	Freer[b] 34.23

[a] Cleveland Museum of Art.
[b] Freer Gallery of Art, Washington, D.C.

can, however, frequently provide important information regarding the source of the ores which were used for smelting.

Sampling is essential in determining chemical composition. Because the final metallurgical structure as well as the concentration of the elements will dictate the homogeneity and, therefore, the possible sampling error, it is important to determine how and if the sampling method inherent in specific nondestructive analytical techniques is affected by metallurgical structure. Thus, we had to use objects from which we could, with the concurrence of the museum curator or owner, remove a small sample for metallographic study as well as for chemical analysis. Since this procedure is not always possible without destroying the aesthetic aspects of the object, the number of samples available for this study is severely limited. Table I lists the objects from which samples were obtained and on which the data generated in this study are based.

Metallurgical Structure

In metal alloys, errors induced in sampling are affected by two different sources of inhomogeneity. The first source is at the atomic level because of segregation in solid solutions. This is usually only a major problem in castings with coring (*1*). The second and major source for inhomogeneity in silver–copper alloys is related to the size and distribution of the phases. This factor is particularly important in multiphase alloys and, in practice, becomes potentially significant in silver–copper when the copper concentration exceeds *ca.* 5.0 wt %. Microscopic inhomogeneity may be the result of equilibrium or nonequilibrium conditions present during any fabrication step as well as the degree of uniformity in distribution of the second phase caused by the fabrication technique.

The accuracy of chemical analysis should be related to the degree of homogeneity of the object as reflected by its metallurgical structure. Thus, the objects sampled have been divided into three categories. Since the metallurgical structure is also related to the copper concentration, classes I through III also represent increasing copper concentration of the alloy and object.

Class I: highest degree of homogeneity; silver–copper solid solution. Objects typical of this structure are the Oval Bowl, King Torso, and the Vicarello Goblet.

Figure 1. Transverse section of sample from flange of joint between halves of the Horse Rhyton (private collection). Illustrates presence of silver–copper eutectic at interfaces between grains of silver-rich solid solution (×225).

Class II: Intermediate degree of homogeneity; small areas of copper-rich eutectic. The structure typical of this class is shown in Figure 1. In some instances the copper-rich phase has a much finer dispersion, and on etching it frequently shows up as a "muddy" diffuse structure at the grain boundaries or over entire grains. Figure 2 shows a typical example of this structure. Objects falling into this class are Anahita Rhyton, wire from the Ewer hinge, both Horse Rhyta, and the Hormizd Hunting Plate.

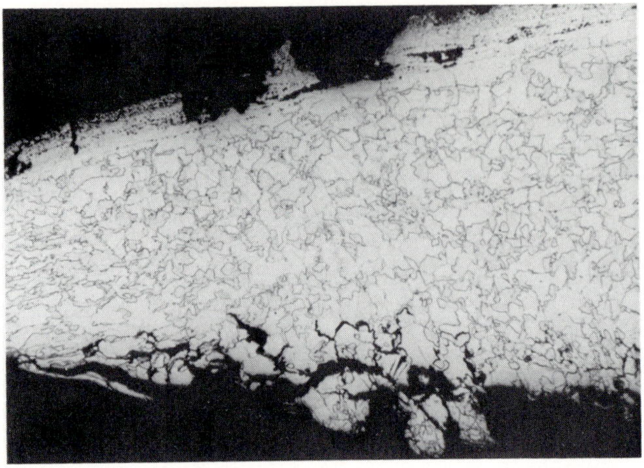

Figure 2. Transverse section of sample from inside neck of the Anahita Rhyton (CMA 62-294). Structure illustrates darker etching grains (muddy), indicating precipitation of copper-rich phase (×55).

Class III: the least homogeneous structure; it is characterized by the presence of interdendritic, copper-rich eutectic. Figure 3 shows the metallurgical structure typical of this class. The Silver Nail and the Bull Bowl are representative of this class.

These three classes are further modified by plastic deformation and annealing when an object is made. Fabrication increases the degree of

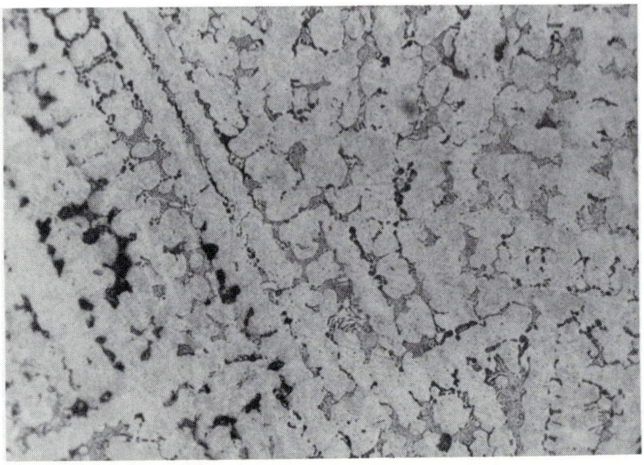

Figure 3. Transverse section from silver nail (private collection). Structure illustrates typical cast dendritic structure. Lighter etching dendrites are silver-rich, and the darker etching interdendritic spaces contain the silver–copper eutectic (×225).

homogeneity. The suffix A after the class designates that the structure has been modified in this respect. Figures 4a and b demonstrate this increasing degree of homogeneity produced by deformation and, therefore, are classified as IIIA. However, the absence of the suffix A does not imply that the alloy was not plastically deformed but only that such deformation, if it occurred, did not refine the metallurgical distribution of the phases.

Homogeneity is further complicated in silver objects of art by the frequency with which the surface was decorated with gilded areas. The measured values for gold concentration, therefore, must be carefully assessed to ensure that they are not related to gilding rather than the ore source or refining technique.

Table II. Analytical Techniques Used

X-Ray Fluorescence

Pt target; 50 keV; 50 ma.

$AgK_{\alpha 1}$ and $CuK_{\alpha 1}$—analyzer crystal LiF.

Detector: GE SPG6 Be window xenon counter.

This technique is insensitive for gold and lead in the range 0.1 to 1.0%, which is the range covered by most silver objects of antiquity and particularly the Sasanian silver objects which are reported here.

Electron Microprobe

Accelerating voltage, 20 keV.

Specimen current, 0.05 μamp as standardized on 100% Ag sample.

Beam diameter, 1μ.

$CuK_{\alpha 1}$ —analyzer crystal LiF

$AuM_{\alpha 1}$ and $PbM_{\alpha 1}$—analyzer crystal PET

Detector: copper—xenon-sealed proportional counter; silver, gold, and lead—flow proportional counter using 10% methane and 90% argon.

Thermal Neutron Activation Analysis

Activation: thermal neutrons; 30 min in flux, 1.8×10^{14} n cm^{-2} sec^{-1}

Characteristic gamma ray energies:

$^{110}Ag^{m}$	(245–d)	658 keV
^{64}Cu	(12.8–h)	511 keV
^{198}Au	(2.8–d)	412 keV

Analyzer Ge (Li-drifted) detector

Analytical Methods

Ultimately, all quantitative analytical methods rely upon standards, whose composition is determined by the classical techniques of wet chemical quantitative analysis. Obviously, the preferred techniques for analyzing art objects are nondestructive, such as x-ray fluorescence, neutron activation, electron microprobe (both dispersive and nondispersive techniques), and so forth. Emission spectrographic analysis is not suitable for major and minor elemental components since it is relatively imprecise in the concentration range > 0.1 wt %. If the counting rate is high enough, x-ray fluorescence can give a precision of $< 1\%$. The area from which data can be obtained on the small samples used in this study results in low counting rates and thus a precision of only $\pm 5\%$.

This study is concerned with the comparison of data from x-ray fluorescence (*2*), electron microprobe (*2*), and thermal neutron activation (*3*) techniques with the metallurgical structure of the sample. The maximum precision which can be expected from all of these techniques, with normal precautions, is $\pm 5\%$. In all cases, the analytical parameters used in the measurements reported here were chosen to maximize the sensitivity and precision for silver-based alloys. The parameters used in each technique are listed in Table II.

Sample Size. Since the aim of this study is to relate the data to the metallurgical structure, all analytical measurements were made on metallographical sections taken from the objects. This procedure limited the volume which could be sampled.

X-ray Fluorescence. The minimum diameter irradiated was 2 mm for accurate data. Some data were obtained with 0.5 mm diameter, but the accuracy is reduced. (This limitation in areas irradiated by the beam is not necessarily true for the general application of this technique to art objects. However, the method samples a surface and thus raises the question as to how representative the volume sampled is of the bulk.) Assuming for silver an average penetration depth of 50 μm (*4*), the volume sampled is $\sim 150 \times 10^6\ \mu m^3 \equiv 1500\ \mu g$.

Electron Microprobe. Since the electron beam diameter is only 1 μm, it is unlikely that the volume sampled represents the bulk composition. To obtain as representative analyses as possible, the focal spot is electronically scanned over a square area of 300 μm $\times$ 300 μm. In some cases, where one dimension of the section was $< 300\ \mu$m, the area scanned was necessarily reduced to 100 μm $\times$ 100 μm. Assuming a penetration depth of 1 μm (*5*), the standard 300 μm square raster samples a volume of $\sim 9 \times 10^4\ \mu m^3 \equiv 1\ \mu$gram.

Thermal Neutron Activation Analysis (NAA). The amount of sample adhering to the quartz sampling plate is variable. It is normally considered to be of the order of 100–200 μg. Because of the size of the metallurgical samples available for this study, the average mass sampled was estimated to be in the range 1–5 μg.

Results and Discussion

Table III shows the copper concentration of each object as determined by each of the three analytical methods. Although the precision of the techniques was ±5% (except for some NAA samples where it was ±10% because of the sample size), the accuracy as a result of sampling and inhomogeneity is much larger. In addition, because good specimen sampling practice is impossible with works of art, the accuracy should be reasonably considered as ~30% for alloys in the solid solution range—*i.e.*, classes I and II, rising to ~100% for class III cast alloys. This is particularly evident from the data on the two samples from the Female Head from the private collection. Sample a was taken from the base of the neck where the metal was thicker and had only been plastically

Table III. Analyses for Copper

	Class	*X-ray Fluorescence*	*Microprobe*	*Neutron Activation*	
Oval Bowl	I	4.8	4.0	4.1	(0.08)[b]
Torso (King)	I	1.7	0.53	0.6	—
Vicarello Goblet	I	—	0.5	0.75	—
Bacchanalian Plate	IIA	—	5.4	3.1	—
Hormizd Plate	IIA	5.8	7.1	4.6	(3.26)
Stroganoff Plate	IIA	3.9	3.9	5.4	(3.87)
Horse Rhyton (CMA)	II	2.5,3.0	2.7	1.6	(1.64) 0.9 surface[a]
Anahita Rhyton	II	4.3	3.7	3.3	(3.31)
Wire (Ewer)	II	5.2	6.4	5.5	—
Horse Rhyton (Private)	II	8.2	11.6	6.0	—
Female Head (Private)	IIIA	(a)8.4	18.5,4.5	13.4	—
		(b)8.1	6.7	6.2	—
King Head (Freer)	IIIA	6.6	0.0,5.2	11.1	(3.32)
Queen Head (Freer)	IIIA	7.4	10.5,5.4,	8.8	(4.74)
Bull Bowl	III	5.8	5.0,5.4, 11.0,16.6	5.9	—
Nail (Private)	III	10.0,10.8	13.0,16.1	7.0	—

[a] Private communication from E. Sayre.
[b] Values in parentheses obtained from A. Gordus (6).

deformed to a small extent (Figure 4a). The structure is quite inhomogeneous, and the copper concentration varied greatly. Sample b was taken from the edge of a break at the back of the head and was deformed to a greater extent (Figure 4b). Here the variation in analysis is considerably less. For the two cast objects which were not deformed, there is again a wide variation in copper concentration as a result of the inhomogeneity of the metallurgical cast structure. This is a particularly

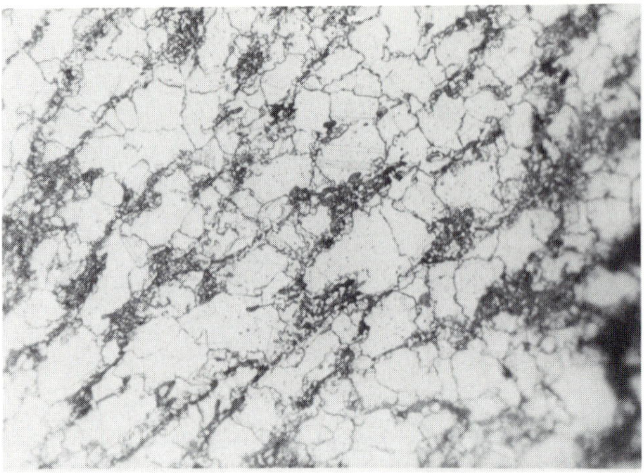

Figure 4a. Transverse section from base of neck of Female Head (private collection). Structure illustrates the slightly deformed cast structure (×225).

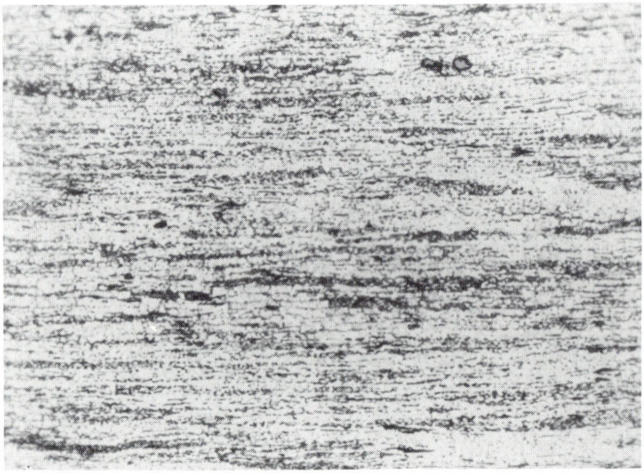

Figure 4b. Transverse section from edge of a fracture in the back of the head of the Female Head (private collection). Structure illustrates the "banded" structure produced by a considerable amount of deformation by hammering (×225).

important observation if composition is used as a basis for discussing authenticity, geographic origin, manufacture, and so forth.

These data also show in conjunction with the data of Gordus (*6*) that the copper concentration as determined from surface rubbings is consistently lower than that obtained from the bulk sample. This result presumably arises from depletion of the copper from the surface either by oxidation during the annealing steps of manufacture or by weathering or cleaning at some time after manufacture.

Table III suggests that x-ray fluorescence is a more reliable analytical tool for determining copper content on as-cast objects—namely, the Bull Bowl and Silver Nail. Presumably, this is simply the effect of the averaging which occurs when a larger volume is sampled.

Table IV summarizes the gold concentration in each object as determined by electron microprobe and neutron activation analysis (NAA). The electron microprobe has a lower sensitivity and, therefore, lower precision than NAA for gold. Despite this, however, the data from the two techniques are remarkably consistent when NAA data are obtained from a recently exposed internal surface of the object—*i.e.*, a metallographic specimen.

Table IV. Analyses for Gold

	Class	*Microprobe*	*Neutron Activation*	
Oval Bowl	I	0.42	0.49	—
Torso (King)	I	0.40	0.45	(0.52)[b]
Vicarello Goblet	I	0.20	0.33	—
Bacchanalian Plate	IIA	0.95,0.52	(5.80)	—
Hormizd Plate	IIA	0.56	(6.9 gilding)	(0.74)
Stroganoff Plate	IIA	0.58	0.76	(0.73)
Horse Rhyton (CMA)	II	0.33	0.20	(0.66)
				0.75 surface[a]
Anahita Rhyton	II	0.40	0.4,0.48	(0.69)
Wire (Ewer)	II	0.13	0.24	—
Horse Rhyton (Private)	II	0.44	0.53	—
Female Head (Private)	IIIA	(a)0.60	0.79	—
		(b)0.17	0.44	—
King Head (Freer)	IIIA	0.25	0.90	(0.78)
Queen Head (Freer)	IIIA	0.37	0.53	(0.66)
Bull Bowl	III	<0.10	0.21	—
Nail (Private)	III	<0.13	0.33	—

[a] Private communication from E. Sayre.
[b] Values in parentheses obtained from A. Gordus (*6*).

The most striking observation from Table IV is that particular care is necessary in interpreting the gold concentration if the object has been gilded over any of its surfaces. This is no problem when the concentration is very high, such as that for the Bacchanalian and Hormizd Plates; here it is evident that despite reasonable precautions some of the gilding was sampled. [The piece from the Bacchanalian Plate (7) was from near a gilded area as was the piece from the Hormizd Plate (8).] With the Horse Rhyton from the Cleveland Museum, it is evident that the area which was sampled by rubbing had originally been gilded even though no external sign of the gilding remained. It is tempting to speculate from these data that both the King and Queen Heads at the Freer Gallery were at one time gilded.

Table V summarizes the data on the lead concentration obtained by electron microprobe analysis of the objects. It was not possible to determine any data on lead from x-ray fluorescence, and NAA cannot be used for lead. The sensitivity of the electron microprobe for lead is poor, and the precision was ±15%. These data suggest that the lead content in the silver objects is consistently low and that lead ore refining was well developed.

Table V. Analyses for Lead

	Class	*Microprobe*
Oval Bowl	I	0.3
Torso (King)	I	0.4
Vicarello Goblet	I	—
Bacchanalian Plate	IIA	0.3(1.0)[a]
Hormizd Plate	IIA	0.2
Stroganoff Plate	IIA	0.2(1.5)[a]
Horse Rhyton (CMA)	II	—
Anahita Rhyton	II	0.1
Wire (Ewer)	II	0.3
Horse Rhyton (Private)	II	0.2
Female Head (Private)	IIIA	(a)0.2 (b)0.1
King Head (Freer)	IIIA	<0.1
Queen Head (Freer)	IIIA	<0.1
Bull Bowl	III	0.2
Nail (Private)	III	0.1

[a] Data from spectrographic analysis (*7*).

The presence of small lead inclusions is easily established by using the x-ray modulation method on the microprobe. We believe that a lineal intercept analysis from pictures taken by this method is probably the most accurate if the electron microprobe is used.

Comparison of these data on the Bacchanalian and Stroganoff Plates at the Freer Gallery with the data obtained on the same objects by Chase (*7*) suggests that the spectrographic technique overestimates the lead concentrations.

Acknowledgments

The authors thank E. Sayre, P. Meyers, and L. Van Zelst of the Brookhaven National Laboratory for carrying out the neutron activation analyses; J. Byrd of the Center for the Study of Materials for his aid with the x-ray fluorescence and electron microprobe analyses; and J. Cooney, Curator of Ancient Art, and S. Lee, Director of the Cleveland Museum of Art, for their help and financial support of this project.

Literature Cited

1. Reed-Hill, R. E., "Physical Metallurgy Principles," 2nd ed., p. 595, Van Nostrand, New York, 1973.
2. Birks, L. S., "Electron Probe Microanalysis," Vol. XVIII of Chemical Analysis, Interscience, New York, 1963.
3. Meyers, P., "Methods of Chemical and Metallurgical Investigation of Ancient Coinage," E. T. Hall, D. M. Metcalf, Eds., p. 183, Royal Numismatic Society, London, 1972.
4. Charles, J. A., Leake, J. A., "Methods of Chemical and Metallurgical Investigation of Ancient Coinage," E. T. Hall, D. M. Metcalf, Eds., p. 212, Royal Numismatic Society, London, 1972.
5. "The Electron Microprobe," T. D. McKinley, K. F. J. Heinrich, D. B. Wittry, Eds., p. 199, Wiley, New York, 1966.
6. Gordus, A., "Application of Scientific Methods in the Analysis of Works of Art," Boston Museum of Fine Arts Symposium, June 1970.
7. Chase, W. T., *Arts Orientalis* (1968) **VII,** 75.
8. Gibbons, D. F., Ruhl, K. C., Shepherd, D., unpublished data.

RECEIVED July 20, 1973.

3

Major and Trace Elements in Sasanian Silver

PIETER MEYERS

The Metropolitan Museum of Art, New York, N.Y. 10028

LAMBERTUS VAN ZELST

The Metropolitan Museum of Art, New York, N.Y. 10028 and Brookhaven National Laboratory, Upton, N.Y. 11973

EDWARD V. SAYRE

Brookhaven National Laboratory, Upton, N.Y. 11973 and Institute of Fine Arts, New York University, New York, N.Y. 10028

Thermal neutron activation analysis was done on small samples from Sasanian silver objects to determine the concentrations of three major elements, Ag, Cu, and Au, and 16 trace elements, Na, K, Sc, Cr, Mn, Fe, Ni, Co, Zn, As, Br, Sn, Sb, Se, Ir, and Hg. Two microsampling techniques (drilling and rubbing) were used for Ag, Cu, and Au. Analyses for major components in rubbing and drilling samples of the same objects showed the effects of surface enrichment for gold and surface depletion for copper. Drilling samples of ca. *500 μg were used to determine trace element concentrations. Multiplicate analyses were satisfactorily reproducible in the silver alloy for Ir, Zn, Se, As, and Sb, less satisfactory for Sn, Sc, Mn, and Br while large inhomogeneities were observed for Cr, Ni, Fe, Co, Hg, Na, and K. We attempted to group objects according to trace element compositions and to relate such groups to art historical information. So far gold and iridium concentrations seem to be the most promising criteria for distinguishing different origins of silver ore.*

The elemental composition of ancient silver objects is a potential source of information on the kind of ores used to produce silver, the location of these ores, the ancient metallurgy used to extract the silver, and the trade routes through which they passed. As part of a comprehensive study on Sasanian silver at the Metropolitan Museum of Art we used thermal neutron activation to analyze small samples from silver objects of the Sasanian period (Persia 227-651 A.D.) as well as from related near eastern civilizations. Combined with information from inscriptions, from stylistic examination, and from studies of other properties of these objects (*e.g.*, method of manufacture, toolmarks) we are investigating to what extent the elemental composition of these objects can yield additional information. Most likely the sources of silver during the Sasanian period were argentiferous lead ores, probably mainly galena (PbS) and possibly cerussite ($PbCO_3$). Because native silver and silver ores rarely occur in the Near East, these sources were much less likely used for silver production (*1, 2, 3*). Silver metal can be obtained from silver-containing galena in two steps. (1) The lead ore is smelted in a furnace under partly oxidizing conditions, and molten lead is collected at the bottom of the furnace; this lead contains the silver as well as most of the metallic impurities. (2) Silver metal is extracted from the lead by cupellation—*i.e.*, the lead is oxidized by heat with air in a porous vessel (cupel). Most of the less noble metals will also be oxidized during this process. The lead oxide (litharge) is absorbed in the porous materials in or near the bottom of the cupel, leaving the silver metal behind. Cupellation can be repeated until the metal has a satisfactory purity. A certain amount of copper, generally about 5%, was deliberately alloyed with the silver thus obtained to increase its strength and malleability.

Impurity concentrations in silver objects depend on:

(1) Initial concentration of each element, relative to the concentration of silver in the ore

(2) Contamination by materials added during smelting and cupellation—*i.e.*, fuel, flux

(3) Evaporation of volatile metals or metal compounds during smelting, cupellation, casting, and annealing

(4) Separation of less noble metals from silver during cupellation

(5) Contamination (for some elements) by the addition of copper

(6) Use of "scrap" silver, especially when obtained from different lead ores

(7) Inhomogeneous distribution of the impurities in both the lead ore and the silver metal.

Since so many factors affect the final composition of ancient silver and since their effects on the final composition are virtually unknown, interpretation of any analytical data could be very difficult. However, previous analyses of silver coins, especially those by Kraay and Emeleus (*4*) and by Gordus (*5*), have shown that the concentration of at least one element —gold—in silver can be indicative of the silver-bearing ore used. Concentration levels of other noble metals relative to silver, such as iridium, can be expected to remain constant or almost constant during silver production. These elements, like gold, could be used to distinguish silver obtained from lead ores of different geographic origins. Other elements might change their concentration levels relative to silver. Nevertheless, these concentrations could still help to "fingerprint" the silver or to yield more information on one or more of the processes used to produce silver.

In this report we describe our quantitative analyses for three major components and for 16 trace elements present from parts per billion (ppb) to parts per million (ppm) in Sasanian silver by thermal neutron activation analysis. The analytical data are discussed element by element. Some of the conclusions which can be drawn from the objects are discussed. Since the amount of comparative data available is too small for generalizations, many of our conclusions and assumptions are subject to change when more data and information become available.

Analytical Techniques

In order not to disfigure valuable ancient silver objects, only minute specimens can be removed for analysis. Sample size is thus a major restriction on the analytical method chosen. Thermal neutron activation analysis provides excellent opportunities for quantitative multi-element analysis of these small samples. The question as to how accurately these samples represent the entire specimen is considered in detail later.

Two microsampling techniques applicable to ancient silver and other metal objects were developed at Brookhaven National Laboratory: (1) surface rubbings on ground quartz supports (also called streaks), and (2) microdrillings that penetrate the interior of the object. These sampling techniques, together with analytical methods for the major components (Ag, Cu, and Au) both in rubbings and drillings and for 16 trace elements (Na, K, Sc, Cr, Mn, Fe, Ni, Co, Zn, As, Br, Sn, Sb, Se, Ir, and Hg) in drillings, have been reported in detail (*6*). Therefore, analytical techniques are described only briefly. Gordus (*5*) also has made extensive use of the rubbing (streak) technique.

The rubbings, either on quartz plates or quartz tubing, each wrapped separately in high purity aluminum foil, were irradiated for 30 min together with rubbings of silver standards of known composition in the Brookhaven High Flux Beam reactor (HFBR) at a flux of 1.8×10^{14} neutrons/cm^2/sec. The gamma-ray energy spectra of each sample were recorded, starting approximately 3 hrs after activation, using an automatic sample changing system and a 40-cc Ge (Li drifted) semiconductor gamma-ray detector coupled to a 3200 or 4096 channel pulse height analyzer. The gamma-ray spectra were computer analyzed for peak energies and peak areas. By comparing peak areas at 412 keV (emitted in the decay of 2.7 d ^{198}Au), 511 keV (12.8 h ^{64}Cu), and 658 keV (245 d ^{110}Ag^{m}) in spectra for samples from objects and from standards, the weight ratios of silver, copper, and gold could be determined. The assumption that these three elements are the only ones present in significant concentrations allows us to calculate the approximate compositions of the silver samples. This assumption eliminates the need to know the sample weight which is difficult to determine for rubbings. Little error is introduced by assuming that silver, copper, and gold add up to 100% since other elements, the major one being lead, are rarely present in total concentrations over 1–2%.

Drilled samples, 200–1000 μg and sealed in pure quartz vials, are activated and analyzed in a similar way for silver, copper, and gold content. Drill bits of tungsten carbide and stainless steel, materials much harder than silver, were used in taking samples. Since no significant differences in concentrations of the major elements of the drill bit materials could be found in samples taken with either of them, possible contamination by the drill bits can be neglected.

The high intensities of the gamma rays emitted in the decay of the radioactive silver, copper, and gold isotopes prevent the observation of gamma rays from most other radioactive isotopes present in irradiated silver samples. Therefore radioactive silver, copper, and gold were separated chemically from virtually all other elements. The procedure for determining 16 trace elements, previously described in detail (*6*), is as follows. Two sets of drilled samples were weighed (200–1000 μg) and sealed individually in small quartz vials. One set, to be analyzed for radionuclides with half-lives shorter than 7 days, was irradiated for 16 hrs at a flux of 1.8×10^{14} neutrons/cm^2/sec in the HFBR at Brookhaven National Laboratory. The second set, to be analyzed for radionuclides with half-lives longer than 7 days, was irradiated for 7 days in the same facility.

Starting 2 hrs after irradiation, samples of the first set were dissolved in nitric acid. Radioactive silver, copper, and gold isotopes were then quantitatively removed from solution by isotope exchange with excess freshly prepared cuprous iodide which was mixed with the solution. Gamma-ray energy spectra of the solutions measured during 100 min with a 40-cc Ge (Li-drifted) semiconductor detector coupled to a 4096 channel pulse height analyzer showed photopeaks of ^{24}Na ($t_{1/2}$ = 15 h, E_γ = 1369 keV), ^{42}K ($t_{1/2}$ = 12.4 h, E_γ = 1524 keV), ^{56}Mn ($t_{1/2}$ = 2.6 h, E_γ = 847 keV), ^{65}Ni ($t_{1/2}$ = 2.6 h, E_γ = 1482 keV), ^{69}Zn^m ($t_{1/2}$ = 13.8 h, E_γ = 439 keV), ^{76}As ($t_{1/2}$ = 26 h, E_γ = 559 keV), ^{82}Br ($t_{1/2}$ = 35 h, E_γ = 776 keV), and ^{122}Sb ($t_{1/2}$ = 2.7 d, E_γ = 564 keV). It was experimentally determined that the separation yield for all of these elements was at least 90%.

The second set of samples was cooled for 10–14 days after irradiation. The silver samples were treated like those of the first set except that silver iodide was used for the isotope exchange. Silver iodide is slightly better than cuprous iodide for this process but does not remove radioactive copper isotopes from solution. However, the half-life of ^{64}Cu, 12.8 hrs, is so short that its activity level has sufficiently decreased during cooling and thus does not interfere with the observation of other radioactive nuclides. The gamma-ray energy spectra of the solutions acquired in 200-min counting periods, showed measurably intense photopeaks of ^{46}Sc ($t_{1/2}$ = 84 d, E_γ = 889 and 1120 keV), ^{51}Cr ($t_{1/2}$ = 28 d, E_γ = 320 keV), ^{59}Fe ($t_{1/2}$ = 45 d, E_γ = 1099 and 1291 keV), ^{60}Co ($t_{1/2}$ = 5.3 y, E_γ = 1173 and 1333 keV), ^{65}Zn ($t_{1/2}$ = 245 d, E_γ = 1115 keV), ^{75}Se ($t_{1/2}$ = 120 d, E_γ = 136 keV), ^{113}In^m (daughter of ^{113}Sn, $t_{1/2}$ = 115 d) (E_γ = 391 keV), ^{124}Sb ($t_{1/2}$ = 60 d, E_γ = 1691 keV), ^{192}Ir ($t_{1/2}$ = 74 d, E_γ = 316 KeV), and ^{203}Hg ($t_{1/2}$ = 47 d, E_γ = 279 keV, corrected for interference from ^{75}Se). Again the quantitative yields of these elements were satisfactory.

Quantitative estimates of the concentration levels of these 16 elements could be obtained by using silver standards containing known amounts of the elements of interest. Two standards were prepared at Brookhaven National Laboratory. Although we have established that the elements in these standards are homogeneously distributed, we have not completed their calibrations. Moreover the standards were not useful for scandium, tin, and iridium. For these elements the concentrations were calculated relative to those of elements for which good standards were available, using nuclear cross-sections and decay constants from the literature (7).

Discussion

Major Components: Silver, Copper, and Gold. Before evaluating any analytical data it is important to study the accuracy and reproducibility of each sampling technique used—*i.e.*, rubbing and drilling. Is a drilling sample of 200–500 μg representative of the entire object? Duplicate analyses on drilling samples from different areas of the same (ancient silver) object gave consistent results for gold (Table I). The variations are within experimental error. Copper analyses show a group standard deviation of the fractional deviations from the mean of 4.5% for copper contents less than 6% and of 15% for copper contents more than 6%. These variations indicate an inhomogeneous distribution of the copper in the silver which can be caused by phase separation which for pure copper–silver alloys starts at about the 5% copper level, poor mixing or local depletion arising from corrosion.

Table I. Percent Group Standard Deviation from the Mean in Multiple Analyses of Drilling Samples

	Cu		*Au*
	<6%	*>6%*	
Number of multiple analyses	11	4	9
Percent standard deviation	4.5	15	2.2

The rubbing technique yields only surface samples. To establish the extent to which these are truly representative of the entire object we compared (1) analyses of rubbings from 30 silver objects, sampled and analyzed by Gordus (8) to (2) analyses of both rubbings and drillings from the same objects, sampled and analyzed by the authors. Because of natural corrosion and cleaning operations, compositional changes at the surface can be expected. Copper (less noble than silver) will be depleted at the surface while gold (more noble) will be enriched. The influence of these surface effects will depend, among other factors, on the depth at which the rubbing is taken. Since surface effects might be strongly related to the fineness of the silver, only objects are considered with a silver content between 90 and 98% which is the most frequently encountered fineness of Sasanian silver. The results of this comparison, expressed in average ratios of the concentrations of gold and copper in the rubbings and the corresponding concentrations in drillings, are shown in Table II. Surface effects are clearly appreciable in the rubbing samples. In the results of Gordus one observes a surface depletion relative to silver on the average of 12% for copper and an enrichment on the average of 12% for gold. For our work these figures are 17% depletion for copper and 22% enrichment for gold. The better results of Gordus can be attributed to a slightly different sampling technique. Gordus cleans the sampling area with emery paper, thereby removing the upper surface layer. We clean this area first using a cotton swab soaked with water and acetone. Then, several rubbings are taken, and the first three or four are discarded. Our less destructive rubbing technique provides samples at less depth which are therefore less representative of the interior.

Table II. Average Ratios of Copper and Gold Concentrations in Rubbings to those in Drillings with their Group Standard Deviation

	Copper Ratios	*Gold Ratios*
Gordus rubbings[a]	0.87 ± 0.21	1.12 ± 0.09
Our rubbings[a]	0.83 ± 0.33	1.22 ± 0.23
Objects sampled	27	30

[a] Relative to our drillings.

Despite an appreciable influence of surface effects on the accuracy of analysis of rubbing samples, this sampling technique is still very useful, especially where a highly accurate analysis is not required or where drilling cannot be permitted. However from silver objects with an internal copper concentration above 10%, rubbing samples often contain only a small fraction of this high copper content. Drilling samples (if possible multiple drillings) are preferred for objects with high copper contents.

Trace Elements. As in the case of the major components, the reproducibility of the analysis was tested for each element by analyzing two or more samples from different areas of the same object. The results, presented in Table III, show for each element the group standard deviation of multiple analyses averaged over a number of objects. The reproducibility is satisfactory for iridium, zinc, selenium, arsenic, and antimony; less satisfactory for tin, scandium, manganese, and bromine; and large variations are observed in duplicate analyses for chromium, nickel, iron, cobalt, mercury, sodium, and potassium. (Although the variations reported here for nickel are relatively small, additional data show that this element should belong in this group.) The large variations for the elements of the last group indicate that they are too inhomogeneously dis-

Table III. Percent Group Standard Deviation from the Mean in Multiple Analyses of Drilling Samples

Element	*Percent Standard Deviation*	*Number of Multiple Analyses*
Ir	11	12
Zn	15	11
Se	19	7
As	14	6
Sb	12	13
Sn	26	4
Sc	37	3
Mn	47	5
Br	48	4
Cr	81	3
Ni	13	2
Fe	63	2
Co	53	7
Hg	61	9
Na	79	5
K	60	5

tributed in the silver for the sampling techniques to be representative. A major cause of poor reproducibility for sodium and possibly potassium is probably contamination. More analyses are necessary for the second group of elements to obtain a more accurate quantitative estimate of the reproducibility of their determination.

Archaeological Implications of the Results

We now have preliminary data on about 70 objects from the collections of the State Hermitage Museum, Leningrad, The Metropolitan Museum of Art, New York and Le Musée du Louvre, Paris. A typical example of a Sassanian silver object is shown in Figure 1. At this point all analytical work is not completed; our silver standards for trace elements have yet to be calibrated. In addition, a proper full interpretation of these results would only be feasible in combination with a critical evaluation of stylistic properties, method of manufacture, metallurgical examination, toolmarks, etc., which does not seem to be appropriate for this volume. Therefore, we confine our interpretation of the analytical data to a few general observations. We will not yet consider those elements which did not yield satisfactory results on duplicate determinations although their concentration levels might be of interest and the concentration ratios between some of them significant.

Many of the objects studied consist of separate parts joined together. For example, most of the plates have a foot which was fabricated as a separate piece from the plate shell. Analyses of foot and plate shell samples can indicate if these parts are made from the same or related silver. Agreement between the two pieces in the concentration of all of the elements which can be determined reliably would indicate a high probability that they were formed from the same batch of silver. If there is a marked difference in the concentrations of those trace elements for which considerable variation is likely during the winning, alloying, and treatment of the metal, but still agreement between gold and iridium which likely retain their concentration levels relative to silver during such processing, it would still be a good indication that the same source of silver was used for the alloys of the two pieces but that the silver was prepared in different batches. Total disagreement, of course, suggests quite different silver.

Of a total of 23 objects for which comparative analyses for different parts are available, seven are matching in all reliable elements determined

Figure 1. Sasanian silver gilt plate showing a king hunting ibexes, late 5th or 6th century A.D. (Metropolitan Museum of Art, Fletcher Fund, 1934)

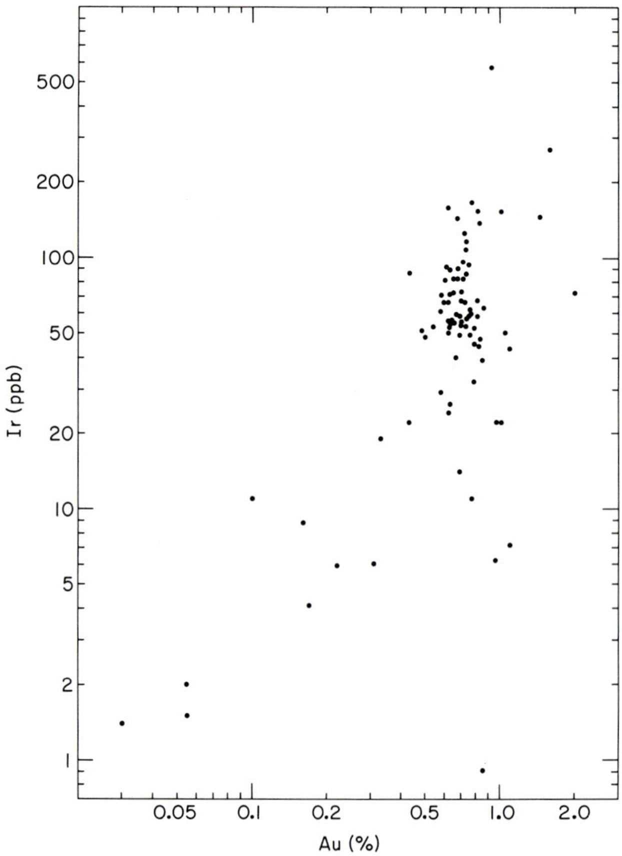

Figure 2. Correlation plot of gold and iridium concentrations relative to silver of all objects analyzed. Analyses of different parts of an object are represented separately.

in the separate parts. These parts, therefore, appear to be from the same batch of silver. Six other objects show matching concentrations of gold and iridium but not of the less noble metals. The separate parts of these objects, therefore, could have been made of silver from a common ore source but not from a single batch. In a surprisingly large number of the objects, 10 in all, there is no match at all in the minor and trace elements. Possible explanations for this disagreement between the separate pieces of these objects (usually the shells of plates and the supporting feet) might be manufacturing tradition, that is, the feet were produced separately as standard items for attachment to plates or that feet were replaced in later repairs.

Because of the extent to which the gold and iridium concentrations seem related to ore sources, objects were grouped according to these concentrations on a log-log scale (Figure 2). A histogram of the iridium concentrations in all of the various specimens analyzed is shown in Figure 3. To avoid the effect of dilution resulting from the deliberate addition of copper, the concentrations have been normalized to proportions relative to silver rather than to total alloy. One observes a scattered occurrence of low gold and low iridium values, a cluster of iridium concentrations between 40 and 100 ppb, and a number of iridium concentrations above 100 ppb—the last two groups with gold values mainly between 0.5 and 1.0%. The objects with low iridium contents, in most cases coupled to a low gold content, are, with a few exceptions, stylistically different from "Central Sasanian" silver objects (7). It is not yet possible to indicate locations of ores used for these objects except that this low iridium group contains mostly objects judged to be from areas east of the "Central Sasanian" area.

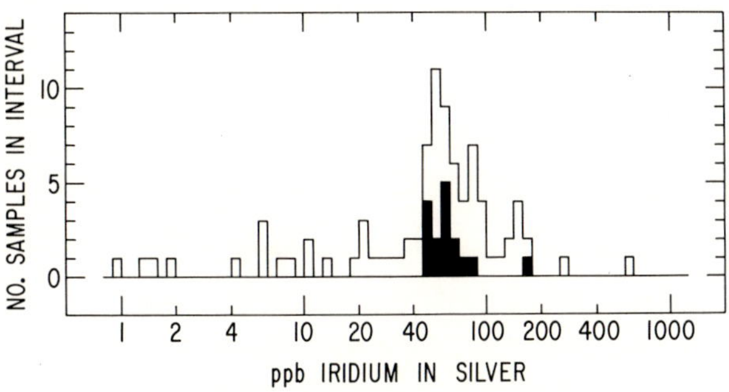

Figure 3. Histogram of iridium concentrations relative to silver in all silver objects analyzed. The distribution of iridium concentrations in shells of plates with external rim lines only are represented by dark columns.

We observed at the back of a number of plates, close to the rim, a chased or engraved line, parallel to the rim. Sometimes a similar line is visible on the foot. These lines do not have an obvious decorative purpose. This group of plates contains objects from a wide time period and various styles. With only one exception all plates with these lines (16 in total) have iridium contents between 45 and 85 ppb (shown in black in Figure 3) while plates without these lines (9 in total) also with one exception have iridium contents either lower than 45 ppb or higher than 85 ppb. It is very unlikely that this correlation is accidental. These similarly lined plates probably had a common origin at least to the extent of having been made from silver from a single source. This particular source does not seem to have been used for the other plates analyzed. Further grouping might be possible on the basis of the analytical data of Sasanian and Umayyad silver coins. Based on the gold content of coins from various mint cities, Gordus (5) could distinguish a number of different silver sources used during the Sasanian and post-Sasanian period. However, the gold content alone does not seem sufficient at this point to subdivide our groups of Sasanian silver objects.

Conclusions

Comparison of multiplicate analysis of micro samples, 200–1000 μg, of carefully homogenized silver standards has established that the analytical techniques developed are acceptably precise for silver, gold, copper, scandium, chromium, manganese, iron, nickel, cobalt, zinc, arsenic, selenium, bromine, tin, antimony, iridium, and mercury. Additional comparison between analyses of similar micro drillings taken from the interior of Sasanian silver objects have shown that such sampling results in data with reasonably good precision for silver, copper, gold, zinc, arsenic, selenium, antimony, and iridium; fair precision for tin, scandium, manganese, and bromine; and poor precision for sodium, potassium, chromium, iron, nickel, cobalt, and mercury. Apparently these last elements are too inhomogeneously distributed in the ancient silver for such microsampling to be representative.

Drilling-sample data from 30 Sasanian silver objects were compared with two sets of rubbing-sample data from the same objects. The rubbing samples consistently have a higher gold content by an average factor of 1.17 and a lower copper content by an average factor of 0.85. Therefore, rubbing samples are undoubtedly biased by surface effects. The extent of this bias, however, does not invalidate the rubbing-sample technique, which is often the only one permissible for certain objects. If the established bias is kept in mind, the results are close enough to the actual interior composition to allow much reliable inference.

Of the 19 elements determined in Sasanian silver objects, only gold and iridium seem to indicate the silver source used. Indirectly through their concentration ratios zinc, tin, arsenic, antimony, and selenium might provide some additional information on the silver source used, and their absolute concentrations, together with those of other elements might be useful in obtaining information on metallurgical techniques used.

Continued analysis of Sasanian silver objects will be directed towards (1) a detailed statistical study of the data, (2) analysis of Sasanian and Umayyad coins, (3) determination of isotope ratios of lead extracted from the silver, (4) correlation between the analytical data and stylistic information obtained from an art historical and archaeological study and with information on methods of manufacture, toolmark, etc.

Acknowledgment

The authors express their gratitude to Prudence O. Harper, associate curator for the Ancient Near East Department of the Metropolitan Museum of Art for her many valuable suggestions, her expert evaluation of the objects analyzed, and her continuous interest in this project.

Literature Cited

1. Wertime, Th. A., *Science* (1968) **159,** 927.
2. Tylecote, R. F., *Metals Mater.* (1970) **4,** 285.
3. Ladame, G., *Schweiz. Mineral. Petrogr. Mitt.* (1945) **25.**
4. Kraay, C. M., Emeleus, V. M., "The Composition of Greek Silver Coins: Analysis by Neutron Activation," Oxford University Press, Oxford, England, 1962.
5. Gordus, A. A., in "Methods of Chemical and Metallurgical Investigation of Ancient Coinage," Royal Numismatic Society *Spec. Publ.* **8,** 127, London (1972).
6. Meyers, P., van Zelst, L., Sayre, E. V., *J. Radioanal. Chem.* (1973) **16,** 67-78.
7. Lederer, C. M., Hollander, Y. M., Perlman, I., "Table of Isotopes," 6th ed., Wiley, New York, 1968.
8. Gordus, A. A., private communication.
9. Harper, P. O., private communication.

RECEIVED July 9, 1973. Research performed partly under the auspices of the U.S. Atomic Energy Commission.

4

Archaeological Ceramic Study Possibilities with a Thermal Gradient Furnace

FREDERICK R. MATSON

Department of Anthropology, The Pennsylvania State University, University Park, Pa. 16802

Changes with increasing firing temperatures in some physical properties of clays sampled near archaeological sites can be determined by firing clay test bars in a thermal gradient furnace and by refiring strips cut from potsherds. A furnace built for this purpose is described, and its possible uses in archaeological ceramic studies are discussed. An example is cited of the changes in color, scratch hardness, and shrinkage of clays now used by Egyptian village potters when test bars were fired in the thermal gradient furnace. The results can be compared with the same physical properties of potsherds. They aid in a better understanding of the ancient Egyptian potters' problems of working with the clays and help to establish criteria for classifying archaeological ceramic materials.

The range of colors found in ancient pottery at one site has frequently led archaeologists to classify their materials into red ware, yellow ware, gray ware, and the like. The colors of the painted or slipped surfaces have also served as classificatory aids. Unfortunately these differences in the fired colors of the clay vessels are technological variables that may have been caused by the time-temperature-atmosphere conditions under which the ware was fired. They do not necessarily have cultural significance which can be used to classify pottery unless distinctly different raw materials were used which fired to different colors. The quantity and size of $CaCO_3$ grains in clays weathered from limestone mountains or hills, soluble salts present in the clays, and the rate of drying of the plastic clayware can all influence the ultimate fired color as much as or even more than the iron content of the clay and the form in which it is present. Slight variations in the clay sources, manufacturing procedures (which can include the mixing together of two or three clays and aplastic), kiln design and operation, and fuels used by the potters in a single pottery making village can result in considerable variation in the appearance of the wares produced. Many archaeologists now recognize that it is necessary to be familiar with the drying and firing properties of the clays available at or near their sites before one attempts to differentiate wares in terms of their body color and the appearance (or actual presence) of a slip or of the painted decoration, let alone identify some pieces as imported.

With the aid of color illustrations Harrell and Russell (*1*) have shown some of the color changes that can occur in several clay bodies when the furnace atmosphere is varied from oxidizing to reducing and when its moisture content is controlled. Brownell (*2*) has summarized the roles of several factors influencing the development of scummed surfaces on brick and other structural clay products. Matson (*3*) has discussed the relationship of the firing temperatures of ancient Near Eastern pottery and the colors developed together with other physical properties of the wares.

Thermal Gradient Furnaces

Thermal gradient furnaces are used in geochemical studies of phase relationships and in ceramic test work, but they have not been used in archaeological studies. There are two basic types of furnaces. In the first, refractory ceramic hollow cores are spirally wound externally with platinum or alloy wires, the spacing of the wire being controlled to produce the temperature gradient desired. This core, packed in an insulated container, is electrically heated with a variable input transformer. Combustion boats containing the samples can be introduced into the furnace. If atmospheric control is desired, a quartz tube can be inserted into the core, the test pieces are inserted, and the gas for the controlled atmosphere is introduced at the cold end. With a removable thermocouple probe the gradient through the length of the furnace can be easily measured. The second furnace is a variant of the first. It is useful for larger test pieces and is a well insulated kiln with all of the heat applied behind the closed end of a horizontal muffle. Such furnaces have been used to survey possible clay sources in the brick industry in terms of color development and degree of vitrification. They can also provide data useful in establishing firing schedules. Both types of furnaces can be used to study the maturation of glazes and overglaze colors. The core-wound furnace is often used in the glass industry for devitrification studies.

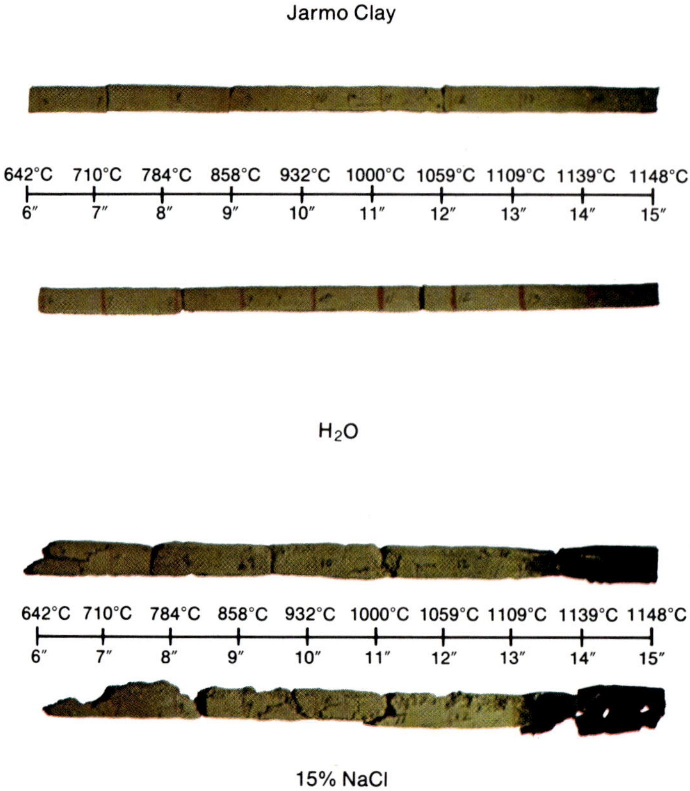

Figure 1. Test bars of clay from Jarmo in northeastern Iraq tempered with plain and with salted (15 wt % NaCl) water

Firing tests of clays collected in the Mediterranean and Near Eastern countries have been done by the author for some years. Briquettes made from these clays were fired to successively higher temperatures with established soaking periods and at times with reducing atmospheres (3). These experiments have provided data concerning changes in color, shrinkage, porosity, and hardness of the clays with respect to firing temperature and time. However, they represent thermal step samplings, and several firings are usually needed to complete a study. With a thermal gradient furnace, using long test bars made from the clays or slices cut from potsherds, variations in kiln temperature conditions closer to those encountered by the ancient potters can be replicated.

Through the cooperation of John Wosinski of the Corning Glass Works, bars made from two Near Eastern clays collected in northern Iraq were, after a preliminary firing to 550°C for 3 hours to remove the chemically combined water in the clay and the initial firing shrinkage, fired in the Corning laboratories in furnaces normally used for devitrification studies. The clays were prepared in two ways to confirm the color changes already established in the step firings of briquettes. One sample of each clay was tempered with tap water; the other was tempered with water containing 15 wt % NaCl. The color changes of the shown in Figure 1. (Some of the earliest pottery yet known was being made at Jarmo before 6000 B.C. The site is west of Sulimaniyah in northeastern Iraq.) The clay was light brown at the lower temperatures, it oxidized to a light red, and then turned pale yellow. At the high temperatures with incipient vitrification a green color developed with the formation of calcium ferrosilicates in the clay body. As would be expected, the salt-enriched clay tended to flocculate and was quite friable. Pale yellow colors appeared at lower temperatures than with the normal clay, and vitrification occurred at the higher temperatures, making it difficult to remove the vitreous samples from the platinum combustion boats.

Furnace for Archaeological Studies

Recently it was possible to construct for archaeological studies a thermal gradient furnace similar to one used in the ceramic laboratories of the College of the Earth and Mineral Sciences of The Pennsylvania State University. Robert Frantz modified the design of the earlier furnace so that it would be suitable for testing archaeological clays and sherds that are available only in small quantities.

The thermal gradient furnace now being used for archaeological ceramic studies contains a 20-inch long SiC muffle that is D-shaped in cross section and is closed at one end. The flat base of the mufflle is 3½ inches wide, and the maximum height of the arch is 2½ inches. Four SiC heating elements 20 inches in length with an 8-inch hot zone are placed horizontally, one above the other, in a C-shaped grouping behind the closed end of the muffle. A variable electric power input 240-V transformer with an output of 0–280 V, 56 amps and 15.7 KVA is used to heat the elements. Their temperature is measured with a Pt–Pt/Rh thermocouple placed between the elements at the closed end of the muffle. This thermocouple is connected to a pyrometric controller so that the rate of temperature rise, if desired, and the holding temperature at the hot end can be readily maintained. The temperature gradient within the muffle is measured by 12 chromel-alumel thermocouples spaced 1⅜ inches apart along the length of the side wall, ⅜ inch above the floor level. A multiple-point recorder is used to print out the record for these 12 couples. The insulated furnace, 19 inches wide, 32 inches long, and 23 inches high, is supported on a wheeled metal cart. The control and indicating instruments, including an ammeter and a voltmeter, are installed on the cart panel beneath the kiln. Figures 2 and 3 show the appearance of the furnace, and Figure 4 shows the interior of the muffle after it has been loaded with test bars.

Many experimental firing schedules can be designed for the furnace. Ideally, the furnace should be maintained at the desired temperatures, but this presents problems of thermal shock when clay test specimens are introduced and removed. A sample firing schedule is shown in Figure 5. The maximum desired temperature (1080°–1100°C) was attained in 3 hours at the hot end of the muffle (thermocouple 12, which is ¼ inch from the end wall). Couple 1 which is 4½ inches within the open end, a plug formed from insulating brick occupying the first 2½ inches, had not yet reached 400°C at the end of the first 3 hours. After an additional 3 hours, the temperatures at thermocouples 8–12 had stabilized. These five couples cover a distance of about 5¼ inches. The temperatures at couples 1–7 continued to rise, but at slower rates after the first 4 hours. The gradient across the 12 couples at the end of the 6-hour firing was

Figure 2. Thermal gradient furnace with control panel

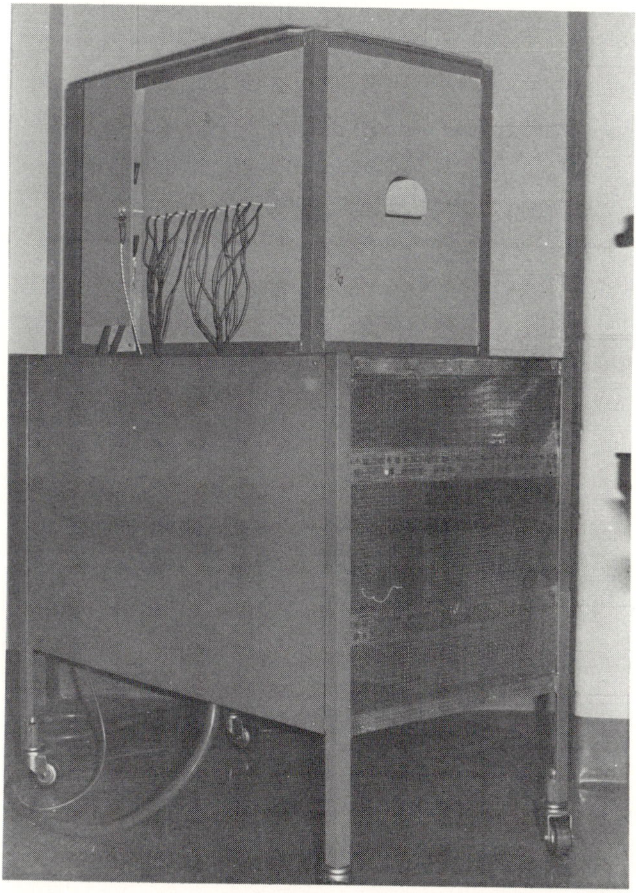

Figure 3. Thermal gradient furnace showing thermocouple distribution

Figure 4. Muffle of thermal gradient furnace loaded with test bars

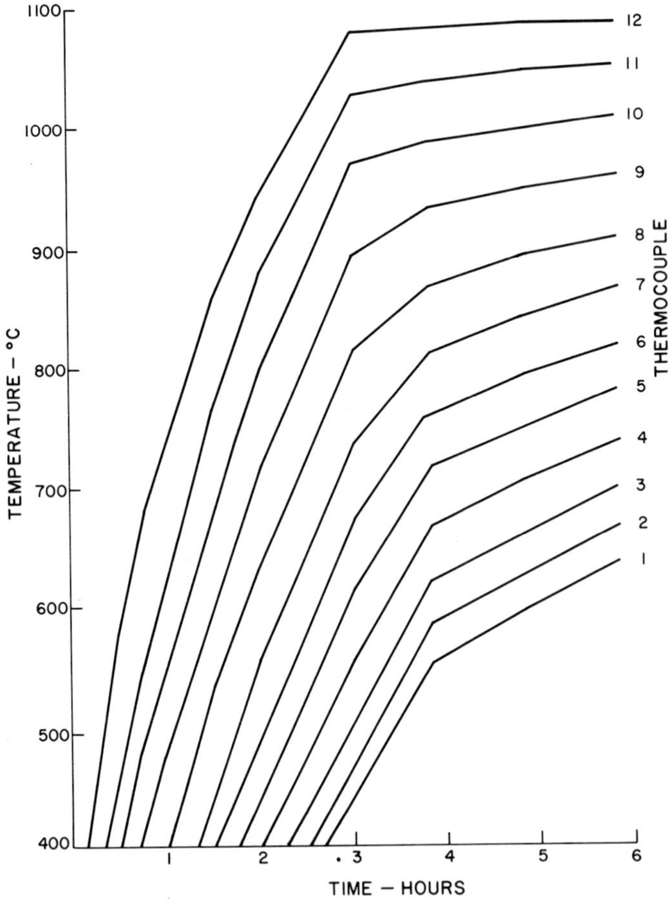

Figure 5. A time–temperature firing schedule of the thermal gradient furnace at the 12 thermocouple points of measurement

475°C over a distance of 15 inches. This provides an ample temperature range to evaluate a clay. If one wishes to have the clay bar at its hot end approach complete vitrification, and there is adequate protection for the muffle, the holding temperature at couple 12 can safely be increased to 1150°C. This firing schedule has been cited to indicate the shortest time in which satisfactory results can be achieved when the kiln is at room temperature at the start of the firing. A slower heating schedule is often more desirable and of course extends the life of the heating elements. The rate of cooling can, if necessary, be controlled as well. If the power is turned off at the end of the 6-hour firing, the temperature in the muffle drops rapidly, and the samples can be removed after 3–4 hours of cooling.

The test pieces for the firing can be prepared in many forms, depending upon the information sought. When studying clays it has been convenient to form bars whose plastic length is 7 inches and whose cross section is no more than ½ inch. Two such bars, placed one behind the other, provide a suitable test strip of clay. They can be placed on a refractory slab or in long combustion boats. Because of the small width of the bars, four clays can be test fired at one time without stacking. Stacking could cause trouble in the hot zone if some of the clays begin to vitrify. One difficulty with long narrow bars is that some tend to warp badly during firing if they are formed from clays that develop high firing shrinkage. For such clays a series of much shorter bars placed in a line one behind the other, or larger bars with a square cross section about 1 inch wide could be used. Strips cut from one large potsherd with a diamond impregnated saw and placed in a gradient line can also be tested in the furnace.

Archaeological Test Problems

When trying to establish the maximum and minimum firing temperatures reached by ancient potters and to estimate the duration of the firings, data obtained from test bars made of the local clays and from strips cut from potsherds, all fired in the thermal gradient furnace, are essential. Careful evaluation must be made of the color, hardness, shrinkage, and porosity measurements, together with a petrographic analysis of mineralogical changes that have occurred if a thorough study is to be conducted. Ideally, a large series of sherds should be studied before the test pieces are selected for refiring, and one should have some knowledge of the variability in the local clay. Since potters working in Mediterranean and Near Eastern villages today do not fire their wares under completely uniform conditions, there is no reason to expect that ancient practices were any different. Therefore it is essential that one have some knowledge of the range in physical variability of the potsherds from each time level in a well-excavated site. Some of the factors that influence the quality and appearance of the village wares produced today are the state of repair of the kiln, the size and positioning of the pieces in the kiln load, the kind and amount of fuel used (seasonal availability and cost), the weather conditions at different times of the year (winds, rain, frost, or heat), the knowledge and skill of the man firing the kiln (usually the potter), and the quality demands of the market for which the ware is produced. The maximum firing temperature in village kilns such as those just discussed varies from about 800° to 1000°C. Occasionally 1050°C is reached but is not maintained. Slagging and slumping result for most earthenware products if yet higher temperatures are attained.

When one is concerned with pottery fired on a hearth in a pile of brush or packed in other organic material such as dung cakes, the maximum temperatures attained for any significant time will be much lower—usually 600°–900°C, and often less. The ancient pottery made by the Indians of the Americas and by peoples in several other parts of the world was fired in this manner. Some Indians of the southwestern United States, for example, continue this practice today. Vessels are at times intentionally blackened at the end of the firing while they are still hot by smothering them in organic materials. Sherds from vessels fired in the ways just mentioned can be usefully refired at the cooler end of the thermal gradient furnace after the original color changes from the surface through the core have been noted for an adequate series of sherds.

Thermocouple probe measurements by Shepard (*6*) of Indian pottery being made in the southwestern United States showed maximum temperatures in the range 625°–940°C, the latter being most exceptional. Pottery made in Papua and New Guinea, according to Lauer (*7*), was fired from 650° to 920°C. The highest temperatures are reached about 20 minutes after the firing begins, but they are not maintained. Total firing time varies from about 16 to 114 minutes.

Some pottery with well-oxidized surfaces that has been either kiln or hearth fired may have dark gray to black cores. While this can result from low temperature short firing treatment, it may also be caused by the too-rapid firing of fine-textured ware (a situation analogous to the black coring of some building brick). While the core color can result entirely from the presence of carbon, black Fe_3O_4 may also be responsible. After such ware is refired, there will often be a residual color difference between the core zone and the surfaces that must be considered when estimating original firing temperatures. Iron in a reduced state, particularly FeO, can exert a fluxing action on the clay body at relatively low temperatures. As a result, sherds after an oxidizing refiring, even to 900°–1000°C, may never develop the optimum color in their cores that is found in the test bars of clay. The surface flashing of the ware arising from flame impingement and other factors in the original firing can also affect the surface color in ways which are difficult to reproduce completely in a test furnace.

The surfaces of ancient pottery were often decorated with a ferruginous slip that could be intentionally fired to a red or a black color. The color change was the result of not only kiln atmosphere and temperature, but at times also of the degree of vitrification of the slip. Color variations in both the painted decoration and in the clay body of sherds from one archaeological site and time are shown in Figure 6. This series of sherds was excavated at Tell Halaf in northeastern Syria and comes from pottery made about 5000 B.C. A complete technological analysis of this ware should start with the study of the variations in color of the body and paint, textural and mineralogical considerations, refiring tests, and if possible, compositional analyses of the paint and body with the aid of a scanning electron microscope. Once the sherds have been studied visually, refiring tests in the thermal gradient furnace can greatly accelerate the analysis and aid in interpreting the appearance of this attractive pottery.

Figure 6. Potsherds from Tell Halaf in northeastern Syria with varying degrees of slip-painted and body oxidation

More than 3000 years after Halaf times the Greek potters perfected the techniques used by the Halafians and accurately controlled the production of black and red slip decorated wares. Some of the possible techniques for preparing slips used in painted decoration such as those found on Greek vases have been discussed by Winter (*4*). The inclusion of potash from leached wood ash in the finely fractionated clay and the manner of firing strongly influenced the appearance of the fired ware.

The rate of cooling of test pieces after firing must be controlled for optimum color to be developed. In cooling the test bars from red-burning Nile mud collected at Qena, Egypt that are seen in Figure 7, the colors visually changed during the last 200°C of cooling as room temperature was approached. Had the cooling been too rapid, the iron present would not have fully oxidized.

The thermal gradient furnace can also be used to check the archaeological significance of the colors of modern building brick, hollow building blocks, and roofing tiles made today in regions where ancient pottery is being excavated and studied. It is easy to obtain samples of the clays now in use, the kilns can be visited, and the color range in the fired

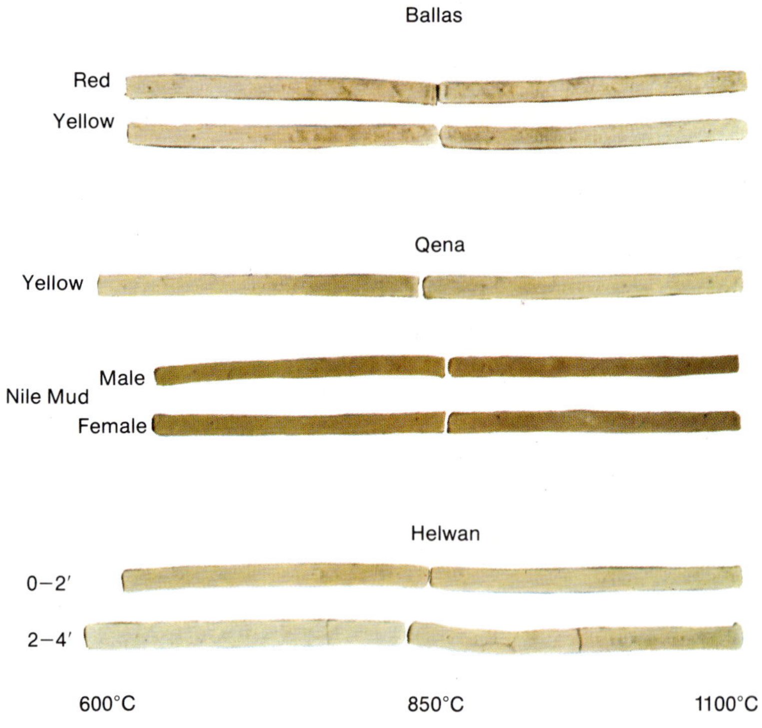

Figure 7. Test bars of Egyptian clays fired in the thermal gradient furnace

products can quickly be established. Figure 8 shows the color variations in a stack of brick fired in a scove kiln near Baghdad. One can observe the variations from tan to red, the yellow-to-white surface scum on some brick, and the conchoidally spalled yellow-green overfired corners of brick that were too near the flames. The thermal gradient testing of clays used

Figure 8. Color variations in contemporary brick formed of Tigris River deposited clay and fired in a scove kiln near Baghdad

for contemporary ceramic products and the study of the industrial wares themselves is desirable, especially when the ancient pottery is limited in quantity or cannot be exported from the country in which it was excavated.

Egyptian Clay Studies

A specific archaeological ceramic problem that has been partly studied using the thermal gradient furnace is that of the variation in the physical properties of the clays used by the village potters of Egypt today—the Nile mud together with clays from the eastern or western deserts. The ancient potters had the same raw materials, and they produced many ceramic wares. By studying the clays it has been possible to determine why the addition of Nile mud to the desert clays improves their working properties and why vessels made of Nile mud alone tend to be coarser. The results have been reported in detail elsewhere (5).

The test bars of the Nile mud were reddish brown (5YR 5/4) in the Munsell color notation at the 600°C zone of the gradient firing, red (2.5YR 5/6) at 850°C, and darkened with incipient vitrification to weak red (2.5YR 4/3) at 1100°C after a holding period of 30 minutes. The scratch hardness, using Mohs' scale, increased from 3.0 to 6.5 for the "male" (better working clay in the potter's terminology), and from 2.3 to 5.0 for the "female" mud. The Nile mud shrank far more when fired than did the other clays—*cf.* the lengths of the test bars in Figure 7—for they were all the same length (7 inches) when formed. The clay from Ballas, north of Luxor, was from the western desert, while that from Qena, also north of Luxor, and from Helwan, just south of Cairo, were obtained east of the Nile. All three areas provided calcarous clays that fired from light reddish brown at the lower temperatures to very pale brown at 1100°C. They had variable degrees of shrinkage, but all were far less than that of the Nile mud. The Ballas clay, an ancient sea bed deposit, developed a hardness of 7 at 1100°C, but the other brown clays had a hardness of only 3. Thus one can readily understand why potters combine clays when preparing the plastic body. The clays used and their proportions, together with the addition of aplastics, varied in the several pottery making villages. An understanding of these physical properties helped in interpreting the fabricating and firing conditions of excavated Egyptian pottery from the same regions; further, they may make it possible to identify wares transported some distance in ancient times from their places of manufacture. Such studies lead to a better understanding of past technological practices and help to establish effective criteria for sorting and classifying the voluminous ceramic yield from Egyptian archaeological excavations.

Conclusion

The selective uses of the thermal gradient furnace in archaeological ceramic studies must be determined through the close collaboration of the archaeologist responsible for the ceramics and the technologist doing the laboratory work. Unless each problem is clearly defined and the archaeological clay samples and potsherds have been acceptably chosen, firing studies can at best be part of preliminary surveys. They should be much more. They should help us to understand the significance of the variations found in pottery excavated and the degree of technological knowledge and control demonstrated by the potters who made the wares as participating members of an ancient community.

Literature Cited

1. Harrell, George O., Ralston Russell, Jr., "Influence of Ambient Atmosphere in Maturation of Structural Clay Products," *Ohio State Univ. Eng. Exp. Sta. Bull.* **204** (1968).
2. Brownell, W. E., "Scum and Its Development on Structural Clay Products," Structural Clay Products Institute, Chicago, Ill., Research Report No. **4** (1955).
3. Matson, Frederick R., "A Study of Temperatures Used in Firing Ancient Mesopotamian Pottery," in "Science and Archaeology," Robert H. Brill, Ed., pp. 65-79, M.I.T. Press, Cambridge, Mass., 1971.

4. Winter, A., "Die Technik des griechischen Töpfers in ihren Grundlagen," in "Technische Beiträge zur Archäologie," Vol. I, pp. 1-45, Römish-Germanisches Zentralmuseum zu Mainz, 1959.
5. Matson, Frederick R., "Technological Studies of Egyptian Pottery—Modern and Ancient," *Cairo Solid State Conf., 2nd, April 1973,* in press.
6. Shepard, Anna O., "Ceramics for the Archaeologist," *Carnegie Inst. Wash., Publ.* (1965) **609,** 77-93.
7. Lauer, Peter K., "Preliminary Report on Ethnoarchaeological Research in the Northwestern Massim, T.P.N.G.," *Asian Perspectives* (1973) **14,** 69-75.

RECEIVED July 20, 1973.

5

Biblical Studies through Activation Analysis of Ancient Pottery

D. BROOKS

Religious Studies Department, Canisius College, Buffalo, N.Y. 14208

A. M. BIEBER, JR.

Department of Anthropology, University of Connecticut, Storrs, Conn. 06268

G. HARBOTTLE and E. V. SAYRE

Brookhaven National Laboratory, Upton, N.Y. 11973

Activation analysis was applied to a group of Persian period ceramics from Tell el-Hesi, Israel. Their elemental compositional patterns, compared with those of clays and pottery from nearby sites, showed that a large fraction of the collection fits the characteristic local pattern. A few of the Hesi sherds matched the clay elemental pattern near Jerusalem. Some of the Hesi materials showed patterns suggesting importation from the Aegean and Cypriote areas. One group of bowl fragments appeared to be imported, possibly from Mesopotamia. Detailed analysis of the results demonstrated the importance of considering correlations between pairs of elements in establishing groups of materials. Tests on modern potters' clays and pottery showed close similarities between the clays and fired pottery.

The technique of neutron activation was applied to ceramic materials from biblical period sites in Israel to aid our understanding of the economic conditions surrounding events mentioned in the Bible. Through trace element analysis, one can often establish the clay origins and hence the probable source of ceramics. A study of the ceramics from a specific site should indicate which were locally made and which were imported from neighboring and distant sites. Trade relationships so established could reflect political conditions in biblical times. Ceramics and clays from Tell el-Hesi and its neighborhood and from sites throughout the Mediterranean area were investigated.

The primary collection of archaeological material, from Tell el-Hesi, was a group of 225 sherds of the Persian period (sixth to fourth century B.C.). A second collection came from Tell Gezer. Both sites are on the coastal plain at the edge of the Shephelah, the low rolling land bordering the central hill country (*see* Figure 7). This region in biblical times was the border between Judah (in the hill country) and Philistia (on the coastal plain) and lay on the major coastal trade route linking Egypt with Asia Minor. Gezer's location gave it control over the major western approach to Jerusalem. Both sites were inhabited from Chalcolithic times (fourth millenium B.C.) down to the Roman period. Tell Gezer is the biblical city of "Gezer," given as a dowry of an Egyptian princess to King Solomon (I Kings 9:6). The identification of Tell el-Hesi is not as certain; most scholars identify it with "Eglon" (*1*).

The archaeological period chosen for investigation was the Persian period, which has received only scant attention. This is partly because of the relative scarcity of archaeological remains, perhaps a consequence of the Assyrian and Babylonian conquests of the land and deportation of important segments of the population. Where Persian remains exist, they have often been partly destroyed by the deep foundations for massive stone architecture erected in the following Hellenistic period (third to first century B.C.).

From the standpoint of biblical scholarship, the Persian period was important in shaping and developing the Judean community from a defeated nation to a religious entity. However, biblical sources for this period are limited to the brief books of Ezra and Nehemiah, some of the minor prophetic books, and an occasional reference or indirect inference in some later work.

Biblical sources imply that Judah with its religious concerns in Jerusalem had favorable support from the Persian rulers. All the surrounding political units which are named—Ashdod, Samaria, Ammon, and Arabia—are depicted as hostile to Judah and its religious interests, and their attempts to undermine Judah, either by direct attacks or by sabotaging its relationship with the Persians, are reported (Nehemiah 4:7, 6:1-9; Ezra 4:4-5). The community also faced an inside enemy—*i.e.*, those members whose contact with the Gentiles caused them to withdraw from the faith of their fathers. Specific reference is made to intermarriage and children who speak the language of Ashdod but not of Judah (Nehemiah 13:23-27; Ezra 9:1-3, 10:1-5). The solution to internal problems that the community adopted, at least as a temporary expedient, was one of isolation from its Gentile neighbors. Since the biblical sources, primarily Ezra and Nehemiah, are written from the isolationist perspective, any independent data concerning the relationships (trade included) of Judah with its neighbors would be useful in evaluating the views presented in the Bible.

Historical sources tell us that Palestine was a part of the western half of the Persian Empire. It belonged to the satrapy "Beyond the River" which included Syria as well (*2*). The satrapy was divided into provinces which, in the region of concern to this project, included Ashdod on the coast (the old state of Philistia) and Samaria (formerly the state of Israel) to the north of Jerusalem. Judah, with its capital at Jerusalem, was also a province. Gaza and the area to the east and south of that city are listed as being in the province of Arabia, but there is conflicting evidence in the literature about the status of Arabia; it may well have been often semi-independent of the Persian Empire. Only the centers of the provinces are named, and the boundaries between them are, for the most part, not delineated. Hesi is in the border area of the Judah of pre-Persian times, and it is about equidistant from Ashdod and Gaza. These facts, plus topographical considerations, indicate that Hesi was probably at a junction of these three political units in Persian times. From the standpoint of the historian, establishing preferential trade patterns would help clarify economic relations in this area and perhaps reflect the political situation.

One archaeological problem which we hoped could be attacked by ceramic analysis studies concerned the meaning of the sunken grain silos at Hesi and other sites in the southern coastal plain (*3*). One theory is that they were supply depots needed for the Persian army in any police action against Egypt (*4*). At Lakhish these grain silos and remains of a large official brick residence are also found. Lakhish may have been a regional center with outposts at small sites such as Hesi (*5*). If this arrangement still existed in the Persian period, did the political ties of the Lakhish complex extend to Ashdod, Arabia, or Judah? The Bible lists Lakhish among the towns of Judah (Nehemiah 11:26), but there are questions about the historical accuracy of this statement.

In the third millenium Tell Hesi covered 37 acres; later, only about 8 acres were occupied. In the Persian period, the ceramics indicate continuous occupation, but there is little structural evidence as to its nature. From the sixth century B.C., we find only occupational surfaces. In the fifth century, a sunken square structure was built; by the fourth century when the characteristic silos were built, it was partly filled in to be used as a dwelling (*6*).

If Hesi were only a grain depot, it may have imported all its pottery; ceramic analytical studies might then tell us by the sites from which the pottery was imported, to which province it was linked. The studies could also show if Hesi had close ties to Lakhish or its neighboring silo sites. If these sites were part of a Persian army supply network, one might expect to find a high incidence of foreign ceramics whose origins could be indicated by analysis.

Experimental Procedures

Sherds from the Persian Period Stratum of Tell el-Hesi. The ceramic materials came from field I, areas 1, 2, 3, 11, 21, 31, in phases A, B, C of stratum V (Persian period). Field procedures called for the excavated sherds to be kept in buckets by loci. The buckets were filled with well water, and the sherds were allowed to soak overnight. Any remaining dirt on the sherds was scrubbed off with nylon brushes and clean well water. Washed sherds were dried in the sun, and representative types were selected to be registered, labelled, and drawn.

The process of producing a complete set of scale line drawings, typing, dating, and conducting comparative research into temporal and typological parallels of the selected material is being carried out by the field I director, W. J. Bennett, Jr. (Associate Professor of Religion, University of Southern California). From his familiarity with the entire group of sherds, Dr. Bennett selected the 225 sherds that form the basis of this activation study. The selection was made to represent each classification, typologically and temporally, of the material found in the designated locations.

The activation materials were identified by two letters (P, indicating the stratum and A, B, C, E, F, or G indicating the areas 1, 2, 3, 11, 21, 31, respectively) and two digits.

Materials from Sites near Tell el-Hesi. During the summer of 1972, over 20 sites near Hesi were surveyed for ceramic materials and clay (*see* Figure 7). At each site sherds were collected from as many time periods as possible. Where available, mudbrick samples were also acquired. Clay samples for each site were taken from wadi beds; specific mines or deposits in fields and limestone clay pockets were also sought. The sites in the plain north of Hesi included Muqanna, Safit, Zippor, and Gat. Collections were made at the two small tells, Qeshet and Sheqef, which are within 2 km of Hesi. The plain sites south of Hesi were Nagila, Milha, Garor, Serai, and Farah. The Shephelah sites included Azeka, Goded, Burna, Maresha, Lakhish, Beit Mirsim, and Halif. Activation analysis results have already been published for Ashdod (*7*). Material was obtained from the Gamma excavations and from the Nagila excavation collection. (The above site names are listed as they occur on the Israel (English) 1:250,000 map, and the reader should be aware that there are several variations on the names of different sites.)

Attempts were made to collect samples of the limestone-associated clays in the hill country. Samples were obtained from the el-Jib area, from near Qastel, at Beit Ummar, and at Arub.

Material from the Gezer excavations had already been obtained. The geologist provided a sample of limestone clay local to Gezer, which however was from a deposit that had not been exposed in antiquity. Previous studies (*8*) have shown that the pottery at Gezer was made from the two main types of clay sources.

Approximately six sherds per site were selected for the initial analysis. Where necessary to establish a group, additional sherds were analyzed. One to three clay samples per site were run. The pottery was treated in the standard manner. The clay was dried for 24 hours at 65°C, ground in an agate mortar, and then treated in the same manner as the pottery.

Materials from Distant Areas. Attempts were made to obtain ceramic materials from major areas where trade to Hesi could have originated. Much of the material for this purpose was borrowed from museum collections. From the Mesopotamian area 39 samples were analyzed. The material was mostly sherds, but there were also two bricks and six cuneiform clay tablets. The material ranged from third millenium B.C. to Islamic. From Iran, nine samples were obtained from sherds from several different areas. The sampling from Syria and Turkey was limited to 12 sherds from the Mersin-Tarsus area and 15 from Syria from the two regions of Aleppo and Hama.

The Cypriote material was extensive because a project similar to that of Hesi was being undertaken at Idalion. A total of 72 sherds, ranging from Late Cypriote (fifteenth to tenth century B.C.) to Roman times, and 10 clays were analyzed. Forty-five samples from sherds from Crete, Mycenae, and Laconia, and some materials from Lesbos, Rhodes, and Aegina were also analyzed.

The needed Egyptian data were obtained from as yet unpublished work on Egyptian clays, Nile muds, and ceramics carried out at Brookhaven by Sami Tobia. Since the results compared favorably with published material (9), further Egyptian studies were not carried out.

With all this material, a standard pottery procedure was followed. Clay samples were ground and then treated like the pottery samples.

Comparison of Modern Pottery with Its Clay Source. Material and information concerning present pottery techniques were obtained in Israel from a potter, Rajab al-Fakhory, working in Hebron. The potter used two clay sources, one associated with a limestone outcropping, which he called yellow clay and traveled some distance to obtain, and the other which he called red clay and took from the field next to his shop. He made some small souvenir vessels from the yellow clay alone, but he usually mixed equal parts of both types of clay. The clays were added to a tank of water, letting the rocks and larger particles settle to the bottom, and then the finer suspended material was removed to another tank. Sand temper (*ca.* 10–20% by volume) and Dead Sea salt (1% by volume; the salt gave a white surface to the fired vessel) were added, and the material was allowed to dry to the proper consistency for throwing.

Samples of the different dry clays were taken, as well as those of the prepared mix, the sand tempers, the salt, and the fired pottery. Samples were also obtained from a neighboring potter who followed the same procedure but obtained his clays from a different location.

Similar material and information was obtained from a potter in Gaza. The Gaza potter did not mix clays but used only the red field clay. The Gaza clay was prepared by sieving the field dirt obtained from the orchard next to the shop, mixing it with water, and allowing it to dry to the right consistency. Some sand temper (about 5%) was added. The potter noted that less sand was used in Gaza than elsewhere because the soil in coastal Gaza contains more sand. The Gaza tradition did not emphasize the white pottery favored in Hebron, but rather black pottery obtained by creating reducing conditions in the kiln during the firing by introducing organic material. Figure 1 shows the various ceramic colors produced by the Gaza and Hebron potters with the clays used there.

In analyzing the material, the pottery was treated by the standard procedure described in detail by Abascal *et al.* (*10*). The clay was dried for 24 hours at 65°C, and approximately 3 grams were ground in an agate mortar and then treated in the same manner as the ground pottery. Some of the prepared clay was used to make small bricks as thick as the average potsherd (~0.5 cm). The bricks first were allowed to dry in air for a week. Some of the dried, but not fired, bricks were drilled and processed by the pottery procedure. The others were fired at specific temperatures (600°, 800°, 1000°C) either in a short time (20 min) or by a procedure in which the kiln was brought up to the specific temperature over a 4-hour period, maintained for 2 hours, and allowed to cool gradually. The fired bricks were then treated by the ceramic procedure.

Figure 1. Palestinian clay types, in lower row, reading right to left, Kaolin (Negev), red field clay, yellow limestone clay. Upper row, four pieces from Gaza, all made from red field clay. Pottery fragment in lower row: Hebron, made from mix of red and yellow clays with salt added to give a white surface to the red paste.

Results

Comparison of Modern Pottery with Its Known Clay Source. The tests on the modern pottery and clay sources from Hebron and Gaza yielded encouraging results; they showed little difference between the clays and the pottery in terms of elemental composition except for a significant increase in sodium content resulting from the addition of Dead

Sea salt to the pottery. More importantly, there were no significant differences between the samples fired at different temperatures or for different times. Raw clay, the water-prepared potter's clay, and the fired pottery showed no meaningful differences. Thus, we simply ground dried clay samples and treated them according to the pottery procedure.

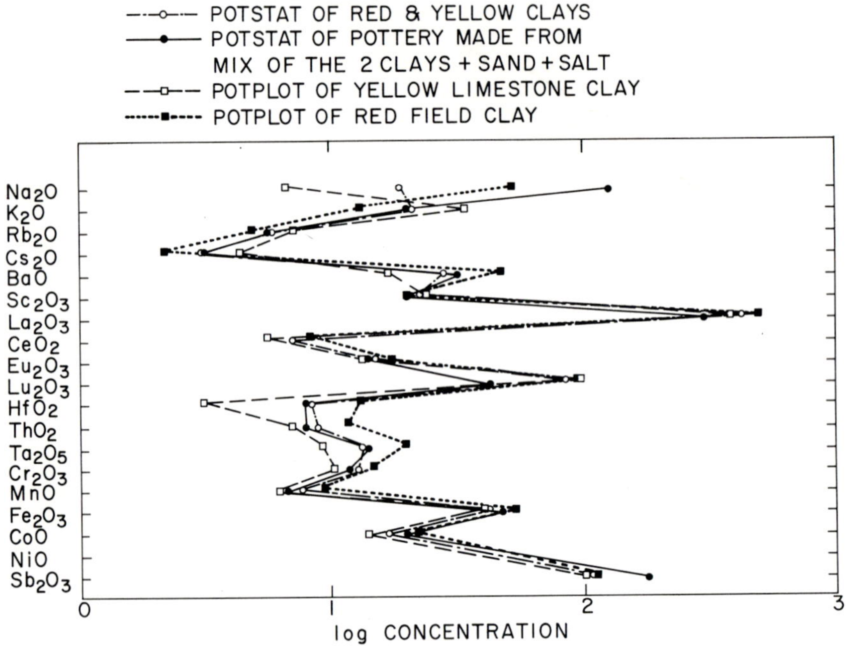

Figure 2. Elemental composition pattern of the red and yellow clays compared with calculated mix of the red and yellow patterns and with pattern of pottery sample made from an actual mix of the two clays plus added sand and Dead Sea salt

When the results of the yellow limestone clay and the red field clay analyses were processed by the POTSTAT routine (*10*), an average of data from the two clays yielded a pattern that matched (except for sodium) the analytical pattern of pottery made of a mix of the two clays. In the first case, the two raw clays were simply ground and analyzed separately; in the second case, the two clays were mixed in a water bath, sand and Dead Sea salt were added, a vessel was formed, dried, and fired, and this finished product was analyzed. The sand temper did not contribute significantly to the relative test element concentrations, but the salt addition did, of course, raise the sodium concentration. These results are graphed in Figure 2.

A separate analysis of three different sands from Gaza, Hesi, and the Hebron area, presented in Table I, showed that the relevant element concentrations are very low; this result led to the conclusion that anything less than about 25% temper should have negligible effect on the relative concentration of elements in the ceramic pattern.

Correlational Analysis of Ancient Ceramics. As we began to apply the ceramic taxonomic techniques (*10*) to group the more than 600 samples from the Aegean, Cyprus, Mesopotamia, and Palestine, we noticed immediately that although there were recognizable subgroupings corresponding to localities, there were also distinct patterns over large areas. For example, the entire Aegean area exhibited to a first approximation one particular pattern, the Palestinian materials had two patterns (one for the red field clays and another for the limestone clays), and the Mesopotamian materials showed still another. There were of course exceptions where some clay sources showed radically different compositional patterns from the regional norm; nevertheless, we regard it as one of the most important conclusions of this study—*i.e.*, in some instances clays from a given large geographical unit can be grouped and identified with that unit through their compositional patterns.

A relative fit matching technique led us directly to a highly significant result when it was applied to the Palestinian clays and pottery. The ceramic group for this area varies a great deal in the purity of the paste and the amount of temper. The clays and mudbrick that were being compared with the pottery also varied in purity. We assumed that the elemental concentrations would be diluted by varying degrees by temper or naturally occurring impurities such as organic matter, but that this could be compensated by using the feature of the POTPLOT procedure (*10*) whereby two plots are slid relative to one another over a light table along the log (concentration) axis. In making these shifts in the plots of the red field clay materials, it was noted that the iron and scandium points on different plots usually could be matched exactly, meaning that the log concentrations of these elements were in a constant linear difference relative to one another—*i.e.*, the iron and scandium had a constant ratio. Such an iron:scandium correlation has often been observed in minerals. Soon the practice developed that when one plot was compared by shifting in relation to another, the first move was to match the scandium and iron points.

Examination of the Egyptian material led to a further discovery. The general Nile pattern, produced by Nile muds and the ceramics made from it, is on first glance very close to that of the Palestinian red field clay. When the plots of three clays and six sherds were matched by the scandium–iron points, cobalt, chromium, and europium also matched exactly. Thus in the Nile samples, concentrations of the five elements are highly mutually correlated. This observation then allowed detection of subtle differences between Palestinian and Egyptian materials. The

Table I. Some Sa[illegible]

	SPECIMEN	K2C	FE2Ø3	NA2C	RB20	CS2C
		PERCENT				
Zif (Hebron Area) Sand	ZF71	.21	.70	200	9.8	1.12
Gaza Sand	GA71	.40	.19	3430	8.6	.14
Hesi Sand	HS71	.46	.58	6670	8.3	.22

Sc/Fe ratios of the two areas matched closely, but ratios taken pairwise among the other three elements (Eu, Cr, and Co) or with iron or scandium were different for the two areas.

The remarkable correlation of iron and scandium observed in a number of potsherds and clays in the Near East has led us to determine this ratio routinely in all our materials and to study its statistical properties. Two examples show the dramatic improvement in the reduction of the standard deviation (expressed as percent) when one considers the Sc/Fe ratio *vs.* the values of the individual elements.

A group of 29 Mesopotamian sherds, bricks, and clay tablets ranging from the fourth millenium B.C. to the Islamic era, from Ur in the south to Ninevah in the north (eight sites) had iron ranging from 4.83 to 8.28% (mean 6.48 ± 14.4%) and scandium from 20.9 to 38.5 ppm (mean 27.42 ± 14.6%). If there were no correlation, one would calculate a Sc/Fe ratio of 4.23 with a compounded standard deviation of ±20.5%. If however the Sc/Fe ratios of the individual samples are calculated and the mean standard deviation of these numbers is taken, the ratio is 4.23 ± 3.3%.

A second example would be a group of 26 sherds assumed to be of local Hesi manufacture. The corresponding figures were iron 4.17 to 7.34% mean 5.55 ± 16.5%, scandium 15.6 to 27.5 ppm mean 20.52 ± 16.3%. The calculated ratio would be 3.697 ± 23%. The standard deviation of 26 calculated Sc/Fe ratios was however only 2.1%. The small standard deviations (3.3 and 2.1% in the Sc/Fe ratio) remaining are close to that predicted for compounding errors caused by analytical technique.

The two above examples illustrate that not only is the Sc/Fe ratio remarkably consistent within a group representing a given area but that it varies significantly from area to area and from clay type to clay type. For example, our data on the Palestinian red and yellow clays show that the former have Sc/Fe = 3.74 ± 0.09 while the latter have Sc/Fe = 5.46 ± 0.29. Of course, in cases where clays of different types were mixed (either by the potter or by natural forces), intermediate Sc/Fe ratios would be obtained.

A few analyses were run on kaolins from Egypt and one kaolin from the Negev in Israel. The kaolins do not seem to exhibit the Sc/Fe corre-

ed for Temper

CONCENTRATIONS OF OXIDES

BAO	SC2O3	LA2O3	CEO2	EU2O3	LU2O3	HFC2	THC2	TA2O5	CR2O3	MNO	CCO
PARTS PER MILLION											
36	2.8	8,65	10,1	,14	0.000	2,41	1,99	,22	15	13	1.0
189	.8	0,00	8,4	,29	0.000	,83	1,03	0.00	7	102	.8
155	2.3	9,84	12,7	,32	0.000	1,28	1,14	,19	17	107	2,4

lation of the other clays. More work and many samples carefully chosen in regard to geographic and geological considerations are needed to clarify this point.

Temper may affect the Sc/Fe ratios if the material used contains these elements in high concentrations and in a ratio much different from the clay. The three sands tested however showed nearly the same Sc/Fe correlation as the red field clays in the region where they were obtained.

Figure 3 shows some typical examples of Sc/Fe correlation and lists values of the corresponding ratios and standard deviations.

Near Eastern Ceramics. The Aegean material from Laconia, Mycenae, and Crete formed an extremely tight compositional pattern; however, in further work we have subdivided these sherd groups in a convincing manner (*11*). The Lesbos material, which consisted of thick, heavily tempered amphora sherds, had a pattern similar to that of the other Aegean materials although specific differences were also noted. Much of the Cypriote and Tarsus, Turkey materials also had patterns similar to the other Aegean groups. One characteristic of the Aegean sherds was a high thorium/hafnium ratio. This ratio, which is very distinct from that of the Palestinian field clay, has already been reported (*7*). Some of the ceramics on Cyprus showed a very different pattern from the general Aegean one. These included some of the earlier wares, such as the Late Cypriote White Slip II (milk bowls), where the Sc/Fe ratios are also very different (*see* Figures 3 and 4).

The Mesopotamian materials, despite their extreme variety of types and time periods, showed a remarkably consistent pattern. Distinctions between the upper and lower Tigris materials do not seem possible, given the elements obtained in the analysis. The ceramics from the one Euphrates site, Babylon, also fit the Tigris pattern. Materials from sites higher along the Euphrates are needed to complete the profile of Mesopotamia.

The Syrian materials fit mostly into the Palestinian coastal plain pattern, but some were more like the Mesopotamian one. The small sampling does not permit more definite conclusions at this time.

Clays and Ceramics from Sites near Hesi. Analyses of the materials from sites neighboring Hesi have been disappointingly uniform. The

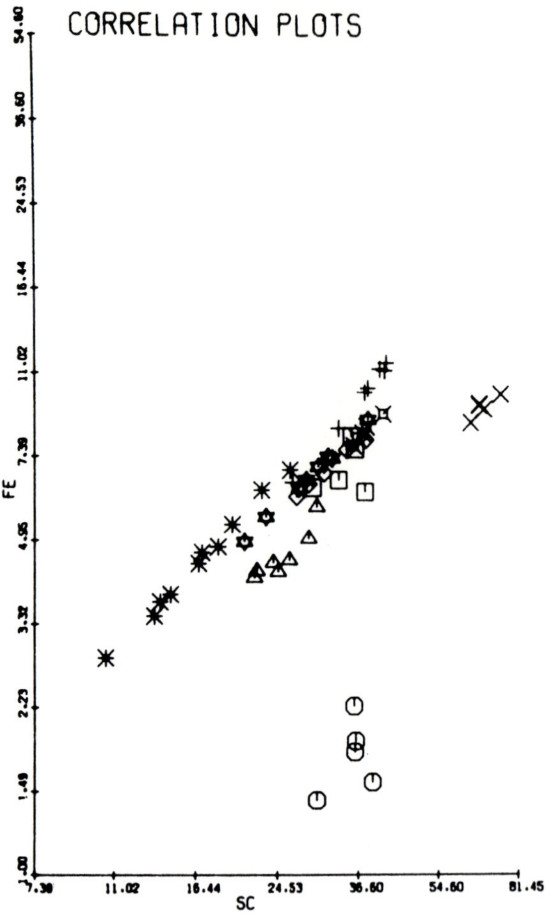

- □ EGYPTIAN LIMESTONE CLAYS SC/FE RATIO 4.94 ± 0.68
- ○ EGYPTIAN & NEGEV KAOLIN SC/FE RATIO 20.23 ± 3.34
- △ PALESTINE LIME CLAY & POT SC/FE RATIO 5.46 ± 0.29
- + NILE MUD & POTS SC/FE RATIO 3.78 ± 0.15
- × CYPRUS MILK BOWL SC/FE RATIO 7.29 ± 0.20
- ◇ MYCENAE SC/FE RATIO 4.52 ± 0.13
- ⌑ HESI BLACK GLAZE SC/FE RATIO 4.52 ± 0.05
- ✳ PALESTINE RED FIELD CLAYS SC/FE RATIO 3.74 ± 0.09
- ✡ MESOPOTAMIA POTS, BRICKS SC/FE RATIO 4.28 ± 0.07

Figure 3. Scandium-iron correlation for various Near Eastern pottery and clay

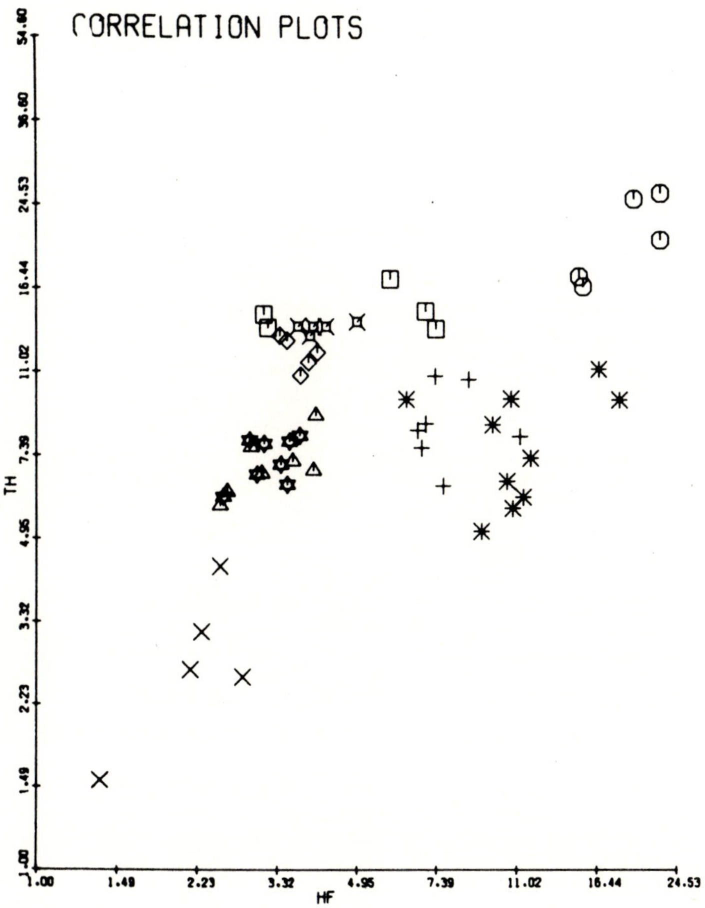

□ EGYPTIAN LIMESTONE CLAYS TH/HF RATIO 3.19 ± 1.28	◇ MYCENAE TH/HF RATIO 3.31 ± 0.47
○ EGYPTIAN & NEGEV KAOLIN TH/HF RATIO 1.13 ± 0.13	⌑ HESI BLACK GLAZE TH/HF RATIO 3.34 ± 0.33
△ PALESTINE LIME CLAY & POTS TH/HF RATIO 2.23 ± 0.29	✳ PALESTINE RED FIELD CLAYS TH/HF RATIO 0.74 ± 0.31
+ NILE MUD & POTS TH/HF RATIO 1.13 ± 0.26	✡ MESOPOTAMIA POTS, BRICKS TH/HF RATIO 2.30 ± 0.28
× CYPRUS MILK BOWL TH/HF RATIO 1.38 ± 0.31	

Figure 4. Thorium-hafnium correlation for various Near Eastern pottery and clay

Table II. Pottery and Clay

Site	Specimen	Type	K_2O	Fe_2O_3	Na_2O	Rb_2O
			PERCENT			
Shephelah and Hebron Area						
Beit Mirsim	BM15	Pottery	1,36	4,74	6160	29,0
Beit Mirsim	BM21	Pottery	1,53	5,75	6320	38,8
Beit Mirsim	BM25	Pottery	1,83	6,72	8260	39,1
Beit Mirsim	BM11	Pottery	1,66	5,68	7640	43,0
Beit Mirsim	BM60	Clay	1,40	5,37	6290	53,6
Beit Mirsim	BM61	Clay	1,23	4,45	6700	40,2
Halif	HL 6	Pottery	1,84	6,65	12100	51,5
Halif	HL 7	Pottery	1,64	6,84	10700	52,8
Halif	HL 1	Pottery	2,47	5,59	8240	46,1
Halif	HL 3	Pottery	1,65	5,25	7460	37,6
Halif	HL60	Clay	1,57	3,45	5640	40,9
Yatta	YT65	Clay	1,35	5,92	4980	48,1
Hebron	HB66	Clay	1,38	5,71	5550	52,5
Lakhish	LK 1	Pottery	1,80	6,10	9520	56,3
Lakhish	LK 2	Pottery	1,58	5,47	7610	53,7
Lakhish	LK 8	Pottery	1,90	5,93	8270	37,5
Lakhish	LK 3	Pottery	2,02	7,41	5290	65,5
Lakhish	LK15	Pottery	1,69	6,64	6520	68,4
Lakhish	LK60	Clay	,91	2,04	3180	24,5
Lakhish	LK67	Clay	1,71	4,61	7580	52,4
Lakhish	LK68	Clay	2,09	4,62	7460	37,9
Azeka	AZ 7	Pottery	1,64	4,24	5580	39,9
Azeka	AZ 8	Pottery	2,36	5,86	6860	56,8
Azeka	AZ 9	Pottery	1,44	5,98	7480	59,0
Azeka	AZ61	Clay	1,27	5,92	6080	57,0
Maresha	MR 1	Pottery	1,75	8,25	6690	76,3
Maresha	MR13	Pottery	1,71	5,32	6900	37,9
Maresha	MR 5	Pottery	1,69	6,03	8210	49,0
Burna	BR 2	Pottery	1,44	6,31	8340	51,8
Burna	BR 5	Pottery	1,51	6,21	6050	45,3
Burna	BR60	Clay	,91	3,29	3790	35,3
Goded	GD 1	Pottery	2,17	6,34	5120	62,7
Goded	GD 2	Pottery	1,88	6,69	4060	66,7
Goded	GD60	Clay	1,19	4,06	4940	45,6
Plain Sites North of Hesi						
Muqanna	MQ 1	Pottery	1,21	5,00	7670	49,8
Muqanna	MQ 6	Pottery	1,27	6,13	9450	57,5
Muqanna	MQ 7	Pottery	1,20	6,10	8840	57,8
Muqanna	MQ 8	Pottery	1,62	6,57	7890	56,0
Muqanna	MQ60	Clay	1,21	4,82	6310	44,0
Safit	SF 3	Pottery	1,44	5,30	7140	45,5
Safit	SF 5	Pottery	1,68	5,97	9270	48,8
Safit	SF 6	Pottery	1,57	5,50	8560	45,3
Safit	SF 4	Pottery	1,42	5,29	6500	51,5
Safit	SF68	Clay	1,57	3,35	7150	36,4
Gat	GT 1	Pottery	1,51	5,72	6400	48,1
Gat	GT 3	Pottery	1,72	6,16	7220	58,3
Gat	GT 4	Pottery	1,70	5,13	7650	46,8
Gat	GT 5	Pottery	2,31	5,49	8090	38,4
Gat	GT68	Clay	2,23	4,18	7950	42,7
Zippor	ZP10	Pottery	2,43	4,92	6670	30,7
Zippor	ZP16	Pottery	1,85	5,10	7570	30,6
Zippor	ZP 6	Pottery	2,58	4,73	6780	31,2
Zippor	ZP11	Pottery	1,95	5,17	7170	28,5
Zippor	ZP60	Clay	2,16	4,68	5480	48,6

Samples from Sites Local to Hesi

CONCENTRATIONS OF OXIDES

CS2O	BAO	SC2O3	LA2O3	CEO2	EU2O3	LU2O3	HFO2	THO2	TA2O5	CR2O3	MNO	COO
				PARTS PER MILLION								
1,04	1560	17.5	36,70	68,3	1.58	.492	8,86	7,76	1.52	125	804	18.7
1,24	2310	21.1	40,60	76,7	1.85	.545	8,54	8,23	1.75	144	907	23.1
1,50	2400	24.6	46,70	83,4	1.95	.610	9,65	8,78	2,05	167	854	27.0
1,84	2100	22.3	46,50	78,9	1.86	.625	7,62	8,65	1.68	280	773	20.5
1,84	440	19.7	38,20	80,6	1.79	.514	10,70	9,76	1.89	144	942	20.4
1,38	478	16.7	35,00	75,4	1.46	.537	16,60	11,30	1.36	142	851	17.3
1,79	740	24.7	42,80	0,0	1.87	.566	13,10	9,31	2.42	186	1230	27,6
1,78	727	25.4	42,70	126,0	2.04	.662	11,50	9,29	2.13	191	1190	30,2
,79	1910	20.4	40,30	0,0	1.70	.542	8,91	8,33	2.15	148	826	31,2
1,38	1100	19.3	41,70	0,0	1.67	.623	8,15	7,64	1.57	140	583	20,6
1,30	464	13.4	25,00	48,9	1.18	.479	11.40	6,09	1.06	120	636	14.1
2,56	478	21.6	56,30	85,0	1.86	1.180	14.40	9,91	2.17	161	1030	23.0
2,40	487	21.4	52,20	91,6	1.91	1.010	14.10	12,70	2.07	157	984	21.8
1,79	540	22.7	45,40	84,9	1.88	.738	9,97	8,95	1,97	153	1070	23.9
1,71	489	19.6	41,50	77,2	1.83	.558	11.10	9,82	1,72	152	1040	21.9
1,29	2300	21.7	38,70	77,1	1.87	.558	9.02	8,11	1,86	149	899	23,1
2,92	894	27.6	60,20	125,0	2.51	.865	13,70	12,60	2,36	223	1840	36.9
2,26	541	23.6	47,50	95,0	2.22	.683	11,50	10,10	2.53	163	1410	30.4
,90	453	8.5	21,10	35,9	,93	.357	6,63	4,38	,70	81	382	7.4
1,70	473	17,2	38,70	72,7	1.59	.632	13,90	8,35	1,47	126	904	18.0
1,17	479	17.2	33,20	72,3	1.51	.523	12,60	10,20	1,46	135	829	18.1
1,48	991	15.9	40,60	0,0	1.48	.604	6,40	6,41	1,42	121	703	15,7
1,84	876	21.3	56,70	125,0	1.83	.589	10,00	9,32	1,97	147	834	22,9
2,01	854	21,7	40,50	123,0	1.89	.663	10,10	8,56	1,87	140	1190	26,3
2,35	452	21.1	40,80	89,2	1,83	.615	13,60	9,96	1,75	153	1040	22,0
2,47	1240	30.2	59,30	114,0	2,58	.966	14,00	12,30	7,49	261	1300	66,3
1,22	1630	19,6	41,50	78,5	1,67	.553	11,20	10,30	1,75	146	807	20,3
1,90	2670	21.9	43,80	87,9	1,85	.602	11,50	9,78	1,96	211	1000	24,3
2,25	1030	24,2	55,80	91,5	2,26	.636	10,30	9,39	2,00	185	1260	28,7
2,03	1120	22.4	46,40	90,5	2,01	.683	10,10	10,20	1,96	183	861	23,3
1,36	484	12,2	25,90	51,0	1,13	.394	9,43	6,23	1,09	118	529	11,8
2,86	690	25,0	50,90	103,0	2,15	.801	13,30	10,40	2,06	217	1360	34,0
2,93	696	25,6	50,60	96,5	2,12	.766	11,70	10,80	2,26	226	1020	32,6
1,75	396	15,2	27,40	57,6	1,36	.511	12,50	7,26	1,30	118	724	15,3
1,74	851	18,9	39,30	78,8	1,64	.605	10,70	8,11	1,86	138	844	22,9
2,25	1290	22.3	46,70	0,0	1,94	.967	9,45	9,08	2.07	151	1150	30,7
2,26	930	22.5	44,90	89,5	1,87	.557	8,82	9,29	2.12	158	999	29,9
2,07	1590	23.7	52,80	0,0	2,04	.724	10,20	9,10	2,18	168	1150	29,3
1,40	478	18,4	33,90	77,3	1,54	.650	18,40	9,75	2,28	167	1060	20,9
2,54	822	19,5	50,60	87,4	1,84	.500	9,74	8,08	1,77	137	917	22,2
2,01	480	22.2	47,80	0,0	2.00	.640	13,90	10,00	2.17	173	1310	35,9
1,34	506	20.7	38,40	98,4	1,87	.634	12,00	9,13	2,08	160	781	26,2
1,73	520	19,1	42,60	0,0	1,73	.583	9,07	9,43	1,76	144	1060	20,8
1,07	440	12.7	26,50	57,2	1,22	.446	11,80	6,56	1,12	85	702	14,2
1,63	674	20.7	47,10	87,6	1,90	.872	8,80	7,90	1,78	140	1010	23,2
1,84	842	22.9	54,00	95,7	2,14	.661	11,10	10,30	1,92	196	813	27,9
1,61	582	18,6	38,80	0,0	1,69	.505	8,99	7,71	1,77	128	996	11,2
,97	1550	20.4	51,20	0,0	1,85	.598	12,40	9,36	1,94	153	921	25,8
1,53	460	15,8	30,70	67,1	1,66	.605	16,80	7,92	1,42	137	820	14,9
1,13	417	18,0	40,30	57,9	1,64	.608	9,86	8,14	1,60	139	815	19,7
1,01	949	18,8	45,90	157,0	1,64	.539	11,30	8,13	1,84	139	812	24,2
1,11	533	17,7	40,30	0,0	1,54	.405	9,62	8,71	1,67	136	728	20,6
,79	377	18,7	43,30	0,0	1,66	.624	9,69	8,55	1,70	136	950	24,6
1,54	388	17,0	29,80	62,8	1,50	.478	11,80	7,37	1,34	126	837	17,7

Table II.

Region	Site	SPECIMEN		K2C	FE2O3	NA2O	RB2O
				PERCENT			
Plain Sites South of Hesi							
	Nagila	NG 1	Pottery	1,97	4,18	5480	33,0
		NG 2	Pottery	1,45	5,21	7000	60,3
		NG11	Pottery	,96	6,30	4190	64,1
		NG13	Pottery	1,77	7,08	5950	46,7
		NG68	Clay	1,26	3,80	6270	36,1
	Milha	ML15	Pottery	1,46	6,26	7360	53,3
		ML16	Pottery	2,10	6,68	5150	76,2
		ML19	Pottery	2,07	6,36	5360	61,9
		ML20	Pottery	1,33	3,31	6390	31,7
		ML21	Pottery	3,36	5,23	2330	84,3
		ML22	Pottery	1,86	5,99	7910	68,5
		ML60	Clay	1,57	3,62	7040	40,4
	Serai	SR 8	Pottery	1,59	5,71	10700	59,0
		SR 9	Pottery	1,50	5,77	9450	51,1
		SR13	Pottery	1,52	6,11	8940	55,3
		SR15	Pottery	1,60	5,95	12900	51,6
		SR68	Clay	1,92	3,70	14300	30,9
	Garor	GR 7	Pottery	1,62	3,98	8170	29,4
		GR 4	Pottery	1,51	5,02	6880	33,5
		GR 5	Pottery	1,86	5,47	6960	34,3
		GR19	Pottery	1,82	4,82	9970	36,2
		GR60	Clay	1,15	2,83	6010	39,0
	Farah	FR 1	Pottery	2,49	6,32	12400	35,2
		FR 2	Pottery	2,07	4,80	12100	29,3
		FR 3	Pottery	2,21	7,31	17100	57,8
		FR68	Clay	1,05	2,61	6370	31,9
Coastal Sites							
	Gaza	GA 1	Pottery	1,46	7,09	11000	63,1
		GA11	Pottery	1,40	6,36	10500	59,4
		GA21	Pottery	1,50	6,55	12600	54,5
		GA66	Clay	1,54	6,97	9560	48,9
	Ashqelon	AQ 3	Pottery	2,13	8,71	15500	107,0
		AQ 4	Pottery	1,25	6,45	13900	53,9
		AQ 9	Pottery	1,33	6,35	9170	53,2
		AQ10	Pottery	2,06	6,28	13200	53,8
		AQ17	Pottery	3,04	5,78	11500	135,0
		AQ60	Clay	1,09	2,62	7460	30,2
Hesi Area							
	Qeshet	QS 6	Pottery	3,01	4,05	3620	55,4
		QS 9	Pottery	1,69	5,33	7540	56,8
		QS 4	Pottery	1,60	5,97	3880	52,5
		QS 1	Pottery	1,98	5,92	9330	51,0
		QS66	Clay	1,46	5,70	7030	49,4
		QS63	Clay	1,29	4,81	9730	49,7
	Sheqef	SQ 1	Pottery	1,75	6,04	5330	41,6
		SQ 2	Pottery	1,80	5,55	5480	47,6
		SQ 3	Pottery	2,01	7,02	11400	55,9
		SQ 4	Pottery	,97	4,19	5070	43,7
		SQ 6	Pottery	1,91	5,96	12000	48,2
		SQ65	Clay	1,35	5,30	7250	45,6
	Hesi	HS87	Clay	1,34	3,31	7290	30,5
		HS88	Clay	1,54	4,69	10600	47,6
		HS89	Clay	,82	1,87	5830	20,7

sherds from each site, with few exceptions, formed tight groups that matched, except for a slightly lower hafnium value, the local field clay and/or mud brick. However, the materials and clays from the different sites could not be significantly differentiated through composition.

The clays taken from the wadis near the tells showed much more variation than the pottery. This may mean that the ancient potter, like

Continued

CONCENTRATIONS OF OXIDES

PARTS PER MILLION

CS2O	BAO	SC2O3	LA2O3	CEO2	EU2O3	LU2O3	HFO2	THO2	TA2O5	CR2O3	MNO	COO
,97	691	15.5	34,30	53,3	1,38	,425	6,71	6,18	1,20	134	581	16,0
1,53	448	19.5	36,80	73,4	1,77	,577	10,90	8,94	1,52	144	913	25,5
2,08	1950	23.5	47,30	93,5	1,93	,572	10,40	10,70	1,85	233	1090	35,7
1,43	1380	25.7	53,90	130,0	2,52	,800	16,90	12,70	2,34	212	2030	48,4
1,07	413	14.1	25,50	55,8	1,40	,468	12,20	6,47	1,25	114	670	14,5
2,00	797	22.7	43,00	77,8	1,95	,565	10,80	8,52	1,94	160	1110	30,4
3,02	832	25.3	57,00	125,0	2,45	,768	15,50	12,20	2,04	210	2960	43,1
2,01	490	24.1	49,90	92,2	2,09	,707	11,30	9,92	2,08	368	1200	26,7
,99	336	12.4	26,40	49,5	1,08	,339	7,65	5,66	1,13	111	588	14,3
5,35	980	28.6	31,00	71,1	1,52	,509	4,60	8,08	1,11	140	973	21,0
1,75	615	22.1	42,70	82,9	1,90	,518	11,50	9,38	1,81	166	1130	26,5
1,29	433	13.8	27,70	57,8	1,21	,562	12,60	7,55	1,19	123	742	13,1
1,53	722	21.4	37,80	0,0	1,86	,434	9,66	8,25	2,13	156	943	34,4
1,42	538	21.1	37,40	85,4	1,88	,512	10,80	8,05	2,14	158	1040	29,5
1,52	660	22.9	40,80	0,0	1,90	,466	11,50	9,49	2,24	167	1150	29,2
1,27	430	22.1	36,20	0,0	1,93	,474	10,50	8,47	2,04	156	903	30,9
1,22	413	13.8	26,80	55,2	1,26	,430	10,50	6,57	1,10	113	613	14,0
,93	920	15.8	28,80	61,1	1,29	,434	8,39	6,36	1,42	140	562	18,5
,85	1030	18.5	43,90	78,3	1,81	,488	9,10	7,75	1,62	167	980	23,3
,94	541	19.8	38,40	87,4	1,56	,487	7,89	9,57	1,80	137	976	24,2
,93	783	17.7	32,70	58,0	1,43	,477	9,54	7,19	1,67	124	815	21,8
,93	395	10.6	20,50	42,9	1,12	,306	9,26	5,16	1,02	76	482	10,8
1,64	4050	24.3	37,10	72,4	2,15	,546	11,60	8,28	2,09	171	870	24,3
,99	317	17.8	32,10	67,8	1,46	,411	8,79	8,03	1,46	127	770	18,1
2,43	585	27.4	40,10	88,4	2,48	,558	9,93	10,30	1,94	200	1240	26,2
,62	284	9.4	14,70	30,8	,85	,351	12,80	3,56	,91	67	453	8,8
2,32	376	26.3	45,50	149,0	2,09	,490	12,40	9,86	2,55	182	1070	43,1
2,33	517	23.4	45,80	0,0	1,88	,634	11,20	9,42	2,22	170	1150	29,2
1,81	443	24.0	39,90	85,6	1,93	,522	10,70	9,37	2,17	179	1240	29,2
1,91	426	26.2	70,30	82,5	1,91	1,090	9,77	8,63	2,07	175	1040	27,4
5,36	248	31.4	41,30	93,9	1,74	,514	4,18	12,00	1,65	1850	1170	66,5
1,40	499	24.4	38,20	81,4	2,06	,572	9,44	9,19	2,35	156	1030	35,7
1,98	454	23.7	44,10	86,7	2,20	,516	9,69	9,51	2,03	152	1560	30,9
1,70	620	22.9	39,90	82,7	2,25	,517	9,05	9,44	1,81	154	1220	28,6
6,44	429	25.4	40,20	87,6	1,63	,500	4,48	14,80	1,53	114	682	18,4
,96	265	9.8	17,60	36,9	,84	,278	8,33	4,15	,73	66	566	9,8
3,52	611	21.8	32,70	56,1	1,37	,396	5,14	6,87	1,00	119	498	15,6
1,36	843	19.7	39,80	75,7	1,77	,532	12,90	8,50	1,58	169	954	22,8
2,10	385	23.1	54,90	95,2	2,19	,572	11,00	10,80	1,81	302	1000	24,6
1,62	610	22.1	41,30	78,5	1,88	,590	11,40	[illegible]	1,70	155	[illegible]	[illegible]
2,00	485	20.0	35,00	75,3	2,56	,496	6,69	9,43	1,44	146	1020	21,7
1,85	354	17.6	31,20	65,8	1,63	,467	8,76	7,46	1,33	124	1010	14,6
1,80	1900	22.9	47,40	89,4	1,95	,529	9,35	9,64	2,57	189	1000	21,8
2,11	655	19.6	41,90	76,8	1,92	,578	8,58	8,89	1,58	195	1120	26,6
1,86	691	26.5	44,30	89,8	2,18	,577	13,00	10,10	2,07	190	1670	31,8
1,25	592	15.3	30,30	56,0	1,43	,416	7,01	6,12	1,14	115	1190	18,8
1,38	847	22.9	48,40	75,9	1,94	,574	11,50	8,40	1,92	182	1480	24,3
2,50	268	20.1	32,70	71,7	1,98	,419	6,41	7,43	1,50	146	558	20,5
1,10	447	13.0	30,70	70,0	1,21	,422	14,20	8,60	1,38	108	591	11,5
1,59	456	16.8	32,60	65,6	1,51	,530	12,60	8,78	1,51	151	959	17,4
,53	376	7.0	15,40	31,3	,67	,252	8,44	4,62	,63	67	345	6,6

his modern Gaza counterpart, used the local field clay, which is really the loess soil, and which covers most of the coastal plain. Analytical results are in Table II. An interesting, similar situation was found at Tell Beit Mirsim. The materials tested from this ancient site fit the pattern of the field clay around the tell. To locate clay sources in this area, we inquired at the neighboring village and learned that when the women of the village need clay (not for potting but for making cupboards, beehives, etc.) they walk to a source 4 km away. Clay from this source was tested and found to be a unique pattern not seen in other clay samples or in any of the pottery analyzed.

The Gezer materials (about 32 sherds) fell into two basic large groups and four other smaller groups. The largest group of 14, which was composed of two subgroups and contained sherds from the total time range tested, had a pattern similar to the sample of local limestone clay. An exact match was not expected since the clay source tested was

Table III. Specimens

		SPECIMEN	K2C (PERCENT)	FE203 (PERCENT)	NA20
Groups of Gezer Sherds Showing Similarity to Local Gezer Clay GZ60	1	GZ01	1,90	4,51	6190
		GZ04	1,92	4,44	5630
		GZ05	1,71	5,07	6040
		GZ19	,92	2,92	2870
		GZ27	1,15	3,74	4240
		GZ30	1,18	3,99	3130
		GZ32	1,27	3,46	3560
	2	GZ14	1,18	4,06	3520
		GZ15	,90	4,08	3440
		GZ24	1,10	3,82	4100
		GZ28	1,48	4,10	4950
		GZ29	1,59	3,99	5500
		GZ31	1,65	4,15	3970
Not Matching Above Groups but Similar to Clay		GZ02	2,97	3,99	4060
Local Clay Sample		GZ60	,34	2,09	1680
Groups of Gezer Sherds Showing Similarity to Coastal Red Field Clay	1	GZ06	1,11	2,51	2310
		GZ11	1,14	2,34	2730
		GZ12	1,55	2,87	3570
	2	GZ13	1,41	5,88	5860
		GZ18	1,42	5,74	7130
		GZ20	1,31	7,00	7820
		GZ23	1,44	5,21	6520
	3	GZ08	1,76	5,38	5790
		GZ17	1,45	5,13	6290
Groups of Gezer Sherds of Unknown Clay Sources	1	GZ03	1,39	3,14	3330
		GZ07	2,10	4,57	4540
		GZ09	2,02	3,96	6820
		GZ22	1,16	4,82	6090
	2	GZ10	,61	2,38	1020
		GZ16	,97	2,20	1450
	3	GZ25	1,78	3,45	5560
		GZ26	2,23	3,05	6820
Gezer Sherd Matching El Jib Clay		GZ21	2,71	4,63	2170

not exposed in antiquity. The second group of nine, with three subgroups dated over the full time range, fit the general red field clay pattern. The remaining nine sherds fell into four small groups; only one had a recognizable pattern—*viz*, that of the El-Jib limestone clay (Table III).

The Hesi Ceramics. As expected, the materials from Tell el-Hesi revealed a large group of sherds with the pattern fitting that of the coastal plain clay. Of the 225 sherds analyzed, 52 belong to this group (Table IV). These were the usual store jars, cookpots, bowls, cups, jugs, etc.,

from Tell Gezer

CONCENTRATIONS OF OXIDES

PARTS PER MILLION

RB2O	CS2C	BAO	SC2O3	LA2O3	CEO2	EU2O3	LU2O3	HFC2	THC2	TA2O5	CR2O3	MNO	CCO
34,9	,99	610	17.8	36,60	65,1	1,69	,561	6,06	6,57	1,34	227	1020	20,9
38,4	1,39	1120	17.9	39,40	66,3	1,69	,524	6,00	6,61	1,16	157	1060	20,8
53,6	2,04	2820	21.6	45,30	72,1	1,77	,559	6,39	8,59	1,51	178	1170	22,8
31,0	1,02	2380	13.0	30,40	45,9	1,35	,399	3,67	5,26	,83	118	0	13,5
33,7	1,52	3150	16.8	33,00	58,2	1,56	,505	5,60	5,70	1,03	169	759	18,7
39,9	2,01	3410	17.9	34,60	58,4	1,54	,547	5,26	6,02	,97	198	968	16,1
30,7	1,11	3080	15,6	33,90	52,3	1,54	,529	5,25	5,35	,93	136	837	15,7
36,9	1,74	1960	18.4	37,50	54,9	1,52	,478	3,90	6,61	1,05	193	837	17,3
35,4	1,48	2350	18,4	37,50	57,7	1,51	,416	3,76	6,57	1,07	167	705	17,7
40,9	1,59	2620	17,7	33,40	55,7	1,52	,506	4,59	6,35	0,00	176	706	30,5
44,2	1,46	2800	19,1	36,70	58,3	1,75	,529	4,00	6,76	1,07	203	601	19,8
38,7	1,55	3150	19,6	38,90	58,6	1,56	,540	3,47	6,72	,98	211	686	17,4
37,5	1,24	3120	18,4	34,10	59,4	1,53	,513	4,94	6,25	1,01	169	803	17,2
48,2	1,62	2580	21.4	40,90	60,4	1,36	,446	1,97	7,15	,97	203	863	18,7
18,3	1,27	3230	10,4	24,80	34,5	1,26	,421	3,50	3,57	,65	96	1180	13,3
22,8	,60	549	9,7	22,10	38,3	,85	,269	3,66	4,46	,81	89	358	9,0
31,2	,65	439	9,5	28,20	38,6	,99	,291	3,88	4,01	,73	83	296	8,3
29,0	,81	246	10,8	23,30	44,8	,92	,308	4,83	4,74	,88	94	490	9,0
46,8	1,39	2480	22.6	44,70	65,2	2,00	,544	9,09	8,58	1,78	180	1270	26,8
40,1	1,49	1520	21.2	48,20	74,9	1,79	,409	9,14	8,03	1,73	199	[illegible]	30,7
59,2	2,07	523	26.0	50,40	0,0	2,27	,618	11,60	10,60	2,32	0	1380	35,7
46,2	1,29	1260	19,8	38,40	73,9	1,63	,472	9,15	8,44	1,49	157	931	25,6
40,1	,93	628	20,6	39,60	82,9	1,79	,496	8,70	8,00	1,55	163	1250	40,9
37,6	,92	1920	19.0	33,40	65,9	1,84	,447	8,86	7,48	1,57	135	948	35,8
26,0	1,05	1150	14,7	37,40	56,2	1,56	,545	6,08	5,28	,85	193	843	13,8
43,2	1,04	359	18,9	39,60	82,2	1,74	,595	8,33	7,06	1,46	273	1170	24,5
40,5	1,26	1900	17,7	41,00	65,5	1,82	,560	9,29	6,83	1,22	190	1210	22,4
41,9	2,63	3300	21,2	46,40	79,3	1,90	,575	12,80	9,11	1,33	215	1100	29,4
9,2	,15	229	9,1	17,40	27,9	,74	,323	,95	2,29	,33	315	132	4,7
14,3	,48	932	10,7	17,50	29,1	,72	,322	1,06	2,80	,32	269	73	4,5
43,7	1,88	2410	17,2	38,90	54,5	1,69	,629	1,85	5,92	,76	271	348	15,1
40,3	1,62	897	15,3	31,60	46,0	1,40	,535	1,88	5,01	,53	193	344	18,9
64,1	2,84	272	21,8	23,80	47,9	1,05	,358	3,33	6,50	,81	124	328	15,5

Table IV. Samples Belonging to

SPECIMEN	K2O	FE2O3	NA2O	RB2O	CS2O	BAO	SC2O3
	PERCENT						
PE29	1.54	4.46	7020	40.4	1.18	883	16.4
PF 3	1.38	5.87	9450	50.9	1.62	550	20.7
PF10	2.14	5.70	9330	36.3	.86	1190	20.2
PF14	1.91	4.73	6800	40.1	1.06	933	18.5
PF21	1.69	4.80	8080	50.0	1.35	941	18.1
PF24	1.82	5.51	8790	50.9	1.52	518	20.2
PF28	2.05	7.34	14100	58.6	2.02	627	27.5
PF31	1.36	6.20	10200	50.7	1.87	520	23.0
PF32	2.21	6.60	9700	41.1	1.72	802	24.2
PF36	1.50	5.21	6710	47.4	1.45	705	19.1
PF39	1.51	5.59	7390	50.1	1.34	3370	20.2
PG 7	1.73	5.14	7980	47.1	1.33	1110	19.1
PG 8	1.61	5.57	9330	40.5	.95	1360	20.4
PG 9	1.74	6.07	9130	54.9	1.67	796	21.9
PG16	1.45	5.39	8230	42.5	1.88	673	19.9
PG28	1.77	6.44	9840	51.4	1.85	811	24.7
PG38	1.62	4.54	8070	32.2	.91	973	16.9
PF41	1.81	5.85	10200	32.9	0.00	828	21.9
PF37	1.46	5.44	7570	44.0	1.34	820	20.3
PF34	1.51	5.22	8340	39.2	1.37	699	19.4
PF20	1.65	7.02	12300	63.8	1.61	566	25.8
PF18	2.00	4.38	8950	33.3	.86	963	16.1
PF13	2.08	4.17	10100	41.8	1.31	700	15.6
PE15	1.50	5.47	7750	33.5	.81	603	20.1
PB34	1.76	5.99	7910	47.5	1.65	524	22.2
PA31	1.77	7.29	12200	60.4	2.23	651	27.0
PC 8	1.45	5.38	9450	50.4	1.58	720	19.9
PE 9	1.39	5.53	7840	47.0	1.60	585	20.4
PE11	1.38	4.34	5470	38.7	1.31	386	15.7
PE20	1.62	5.37	8430	50.8	0.00	535	19.5
PE23	1.56	5.16	7960	43.8	1.07	1330	19.0
PE25	1.75	6.06	10000	74.3	1.66	1040	22.5
PA40	1.55	5.94	9560	56.4	1.79	775	21.7
PB 4	1.87	5.62	8050	34.9	1.20	1120	20.5
PB 6	1.56	4.87	8010	43.0	1.15	1020	18.7
PB19	1.72	5.42	9030	36.2	.98	1300	21.0
PB21	1.98	6.36	10400	55.9	1.80	740	24.5
PB28	1.41	4.10	8190	36.2	1.22	944	15.6
PB33	1.82	5.67	11500	45.0	1.55	2770	22.6
PB36	1.40	5.76	9530	48.9	1.82	849	21.7
PC 3	2.00	5.74	11400	41.2	1.17	2820	20.9
PC 5	1.61	4.80	7180	38.3	.80	540	17.9
PA 8	1.93	5.06	8660	37.2	1.16	1400	18.0
PA12	1.83	5.63	8600	41.3	1.15	1080	20.6
PA15	1.46	4.98	7160	39.5	.71	1110	18.1
PA16	1.87	5.08	9230	49.2	1.29	741	18.2
PA19	1.52	5.54	8190	43.9	1.08	1300	20.1
PA25	2.05	6.17	9740	55.7	1.66	585	22.8
PA27	2.37	4.89	9970	46.0	1.68	1240	17.6
PA34	2.08	6.15	8840	57.9	1.67	716	22.5
PA38	1.45	4.07	7730	35.5	1.49	1270	15.9
PA39	1.89	6.69	11400	56.3	1.41	1780	24.4

and are dated archaeologically to the late Iron II (seventh century B.C.) or the Persian period with a few pieces belonging to other periods. Whether or not the subdivision of this group will be archaeologically meaningful will probably depend on the results of research aimed at the subdivision of the coastal field clays into smaller geographic units.

The next largest group was that of the Black Glaze (actually a shiny slip) ware, also known as Attic ware (Figure 5). The pattern for this

the Local Hesi Elemental Pattern

CONCENTRATIONS OF OXIDES

LA2O3	CEO2	EU2O3	LU2O3	HFO2	THO2	TA2O5	CR2O3	MNO	CCO
	PARTS PER MILLION								
27,80	60,3	1,36	.342	8,33	6,44	1,56	121	741	19.1
36,10	80,2	1,95	.628	10,10	8,82	1,99	147	981	27.4
37,30	78,9	1,74	.530	10,60	8,67	1,87	145	868	21.5
30,50	69,3	1,47	.474	8,43	7,11	1,49	118	713	19.3
30,80	68,5	1,44	.441	10,20	7,84	1,75	142	918	18.8
37,30	81,0	1,51	.562	10,80	8,74	1,77	140	881	21.2
44,20	82,7	1,94	.556	11,90	8,83	2,12	184	1150	28.2
44,00	83,4	1,95	.528	10,10	9,23	2,06	176	1060	30.9
38,00	79,6	1,98	.625	10,10	8,51	1,87	163	870	23.0
35,70	71,5	1,70	.468	8,85	7,66	1,80	132	1010	21.7
39,50	83,0	1,65	.606	9,21	8,45	1,54	161	1210	22.1
33,80	73,4	1,72	.481	10,50	8,05	1,41	136	940	21.8
34,60	78,6	2,00	.585	10,20	8,43	1,75	158	833	21.3
39,60	80,3	1,87	.532	11,10	10,10	2,00	144	987	23.7
35,10	75,2	1,72	.506	8,73	8,34	1,86	150	815	22.3
37,30	81,3	2,11	.562	11,40	9,23	1,75	179	958	25.9
31,50	61,8	1,42	.397	8,90	6,90	1,80	127	614	17.4
36,70	75,6	1,70	.622	11,30	10,30	0,00	157	1120	25.0
34,70	73,8	1,74	.498	9,62	,49	1,80	141	997	25.5
32,00	71,1	1,61	.433	9,04	7,76	1,72	128	921	22.8
38,50	68,1	2,04	.657	9,45	8,96	2,12	193	1140	28.5
28,20	63,2	1,18	.463	8,65	6,83	1,59	118	853	20.1
26,60	63,6	1,33	.495	9,09	6,94	1,56	119	733	15.0
34,40	71,6	1,72	.520	10,40	8,04	1,25	136	886	20.9
34,10	74,6	1,77	.514	8,60	7,91	1,88	142	1070	24.0
39,70	88,1	2,08	.618	10,40	8,80	2,25	178	1160	30.8
32,70	72,0	1,58	.423	9,70	8,15	1,98	161	924	20.2
36,20	73,3	1,71	.447	10,00	7,82	1,75	150	900	22.6
26,90	54,3	1,35	.346	6,83	6,17	1,35	119	860	16.5
34,20	73,7	1,65	.497	9,24	8,76	0,00	137	893	28.6
30,80	67,2	1,51	.404	8,55	7,61	1,57	146	726	20.0
38,10	83,3	1,95	.601	12,40	9,13	,71	164	1200	22.7
36,90	77,0	1,73	.448	8,98	8,11	1,97	165	824	22.8
35,80	81,5	1,61	.640	9,00	8,65	1,50	145	803	19.7
34,10	72,3	1,50	.439	10,70	8,77	1,69	127	877	19.5
35,30	71,1	1,68	.480	9,30	8,55	1,79	157	893	28.9
38,00	64,3	1,86	.546	13,10	9,60	2,08	178	1030	27.0
29,60	58,9	1,36	.426	8,99	7,10	1,51	117	683	15.9
35,20	76,1	1,84	.485	10,50	8,26	1,88	153	760	20.9
41,30	81,6	1,84	.491	11,10	9,22	2,03	151	973	23.1
35,70	81,0	1,98	.609	11,10	9,44	1,71	154	845	19.3
[illegible]	[illegible]	1,50	.408	[illegible]	7,40	[illegible]	143	[illegible]	[illegible]
32,90	73,3	1,68	.568	9,04	8,53	1,45	139	719	17.9
33,40	77,3	1,86	.501	9,96	8,17	1,88	163	1070	26.0
29,80	65,2	1,64	.463	7,17	6,85	1,56	132	731	19.7
31,20	70,6	1,41	.446	8,64	7,75	1,76	136	725	17.9
33,30	74,5	1,79	.546	9,60	8,02	1,91	147	866	22.7
38,40	81,9	1,98	.546	9,45	9,39	1,83	167	1090	24.9
33,60	68,5	1,52	.498	9,53	8,20	1,57	144	1060	18.0
39,50	82,9	2,00	.551	11,20	9,18	1,96	165	1050	24.1
27,10	58,3	1,40	.393	8,90	7,54	1,46	117	619	15.9
37,20	81,0	1,65	.481	9,88	8,38	2,06	168	970	27.1

group of 25 members is very similar to the patterns for Mycenae and Laconia but has a higher chromium content. The pattern closely fits that of a similar group of sherds excavated at Idalion on Cyprus. These two matching groups are assumed to be from Attica, on the basis of their inclusion of one Red-Figured and two Black-Figured sherds. Statistically the number of sherds in this group is not representative of the number of this type found on the site. An effort was made to see if any of this ware could have been the work of an imported potter using local clay; no such evidence was found. Three sherds with black "glaze" deviated slightly from the pattern; two had a yellow rather than the usual orange paste. There were also four pieces (two described by the archaeologists as amphora fragments and two as Cypriote ware) that were also very close to the Black Glaze pattern.

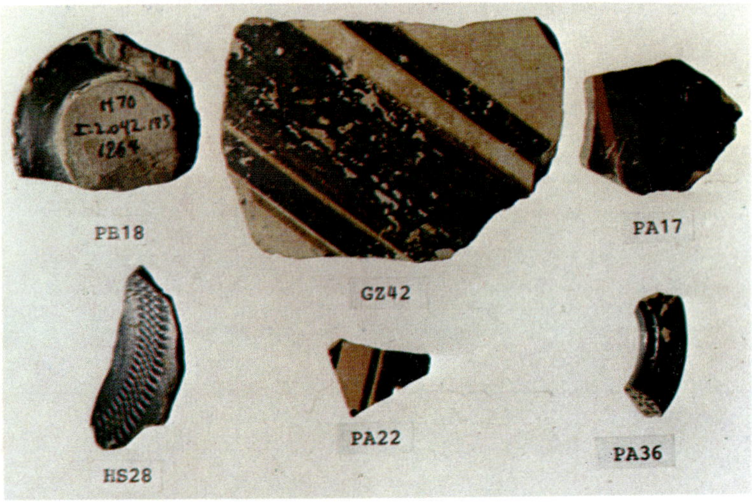

Figure 5. Black glaze ware from Hesi fitting the same Aegean pattern, except GZ42 which is a yellow paste with elemental pattern slightly different from the Aegean pattern

Two other groups had patterns similar to the Aegean pattern. One includes the only whole piece of pottery analyzed, a small juglet. This group (Aegean-like 2) has two sherds described as Cypriote ware. The other group (Aegean-like 3) of slightly differing patterns, is composed of pieces, the pastes of which appear identical to the eye. The second group contains five sherds from the Hesi corpus and one from Tell Burna in the local tell survey.

Possibly belonging to the Aegean pottery was a group of seven sherds (Aegean-like 4) that had a higher Sc/Eu ratio than in any of the Palestinian clays. The ratio was similar to that of the Aegean, but the rest of the pattern, especially in the Th/Hf ratios, did not appear Aegean. Three members of this group were described by the archaeologists as belonging to Greek amphorae; they probably represent an Aegean clay source we have not yet identified. Analyses of the Aegean-like wares are in Table V. A more complete discussion of the Aegean materials is in our report in this area (*11*).

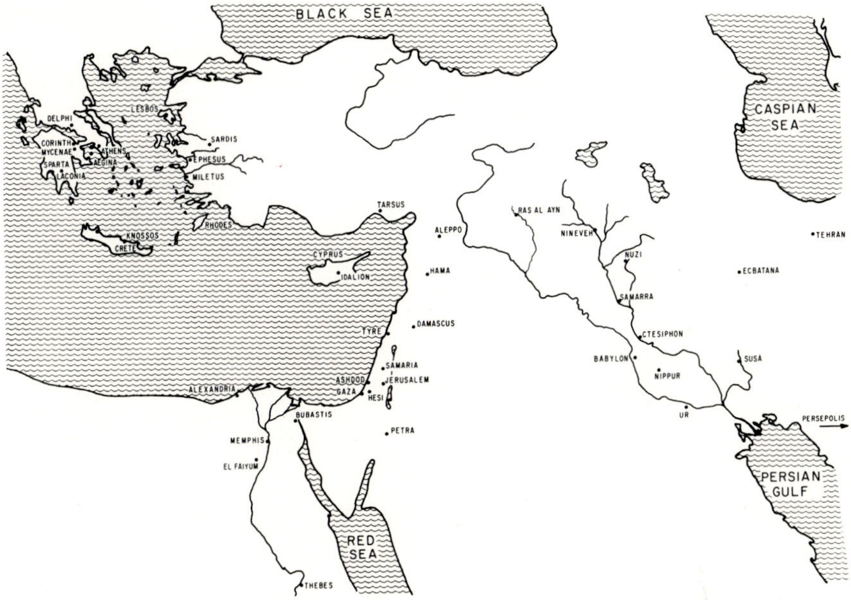

Figure 6. Map of Near East including some of the sites named in this report

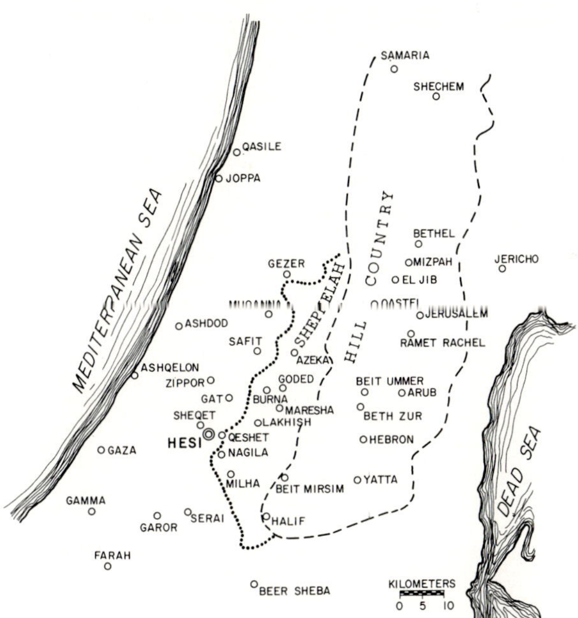

Figure 7. Archaeological sites and clay sources relevant to this research

Historical and biblical students are concerned with trade between plain sites, as Hesi, with Judean hill sites, especially those of the Jerusalem area. At Hesi we found 12 sherds matching the hill limestone clay patterns (Table VI). Also in the local tell survey, two such sherds were found at Goded and one at Azeka. The archaeological description of this

Table V. Samples with

	SPECIMEN		K2C	FE203	NA2C	RB20
			PERCENT			
	HS57		3,19	8,14	6980	163,0
	HS59		3,24	7,83	7890	159,0
	HS60		3,25	8,39	6640	167,0
	PA17		3,59	8,01	9840	155,0
	PA20		3,73	8,58	7270	175,0
	PA22		3,63	8,47	7140	170,0
	PA36		2,89	8,14	7790	161,0
	PB18		3,54	8,29	7300	161,0
	PB32		2,80	8,08	6970	144,0
	PC 2		2,74	8,45	7420	162,0
	PC 4		3,05	8,24	7500	166,0
	PE 1		3,25	8,57	7610	155,0
Hesi Black Glaze	PE 2		3,65	8,81	7280	167,0
	PE 3		3,34	8,30	8690	151,0
	PE 5		3,26	8,31	7890	156,0
	PE 6		3,61	8,46	9120	166,0
	PE 7		3,32	9,11	8790	175,0
	PE22		3,27	8,24	6760	177,0
	PE27		2,98	7,86	7320	154,0
	PF 8		3,38	8,45	8080	166,0
	PG21		3,70	8,83	7790	180,0
	PG32		3,49	8,85	7110	142,0
	HS28		0,00	7,80	0	199,0
	HS30		0,00	8,60	0	216,0
	HS56		3,38	8,62	7410	165,0
	HS62	Black Glaze on Yellow Paste	1,45	8,49	12600	85,2
Hesi Sherds of a Pattern Similar to Black Glaze	GZ42	Black Glaze on Yellow Paste	2,89	6,16	6640	164,0
	PC11	Cypriote Ware	3,57	8,25	7650	164,0
	PC12	Cypriote Ware	3,45	8,93	7060	191,0
	PB 1		2,87	7,85	11200	15,8
	PB 9		2,61	7,94	11100	109,0
	HS27		0,00	8,52	0	177,0
	HS61		3,39	6,06	7670	149,0
Aegean-like 2	PF 6		3,21	6,29	8220	161,0
	PE 4		3,47	7,01	7970	170,0
	PC17		3,09	7,02	7600	119,0
	PF 7		2,03	3,91	2720	72,6
	PA 7		1,78	3,94	2260	65,6
Aegean-like 3	PC16		2,58	5,17	3590	76,5
	PA41		2,35	4,76	2550	76,0
	PC 9		2,20	4,58	3490	77,4
	BR 3	Tell Burna	1,84	5,01	2110	90,0
	PF12		1,32	5,24	4690	38,9
	PA33		3,49	6,12	4980	62,9
	PB 2	Amphora Base	1,39	3,61	4450	36,3
Aegean-like 4	PB27		2,84	5,10	5200	61,1
	PF16		1,20	4,30	4980	35,4
	PG 5		1,37	3,08	4320	0,0
	PG 1	Amphora Base	1,31	4,06	5200	31,5

group has not been completed, except for one sherd described as a storage jar. The hill patterns vary somewhat, and these sherds have not been pinpointed as to specific site.

The literature describes in Palestine during the Persian period fragments of large bowls (20–30 cm in diameter) named mortoria. On the

Aegean-Like Patterns

CONCENTRATIONS OF OXIDES

CS2C	BAO	SC2O3	LA2O3	CEO2	EU2O3	LU2O3	HFC2	THC2	TA2O5	CR2O3	MNO	CCO
				PARTS	PER	MILLION						
12,60	739	36.5	44,50	119,0	1,67	,536	3,91	13,20	1,46	726	1170	48,4
10,70	753	35.9	47,40	137,0	1,80	,441	4,24	13,80	1,63	634	922	60,8
13,40	781	37.8	42,00	59,4	1,87	,460	3,45	13,50	1,44	764	1220	48,1
10,20	667	37,3	41,30	95,1	1,58	,556	4,05	12,90	1,29	676	951	40,7
13,40	841	38,6	39,70	90,3	1,73	,614	4,10	13,20	1,34	816	1160	46,4
13,70	891	37,8	39,40	93,6	1,74	,534	4,38	13,50	1,32	801	1200	46,0
11,00	574	36.6	41,70	94,2	1,80	,514	3,98	13,70	1,43	737	1130	47,2
13,10	693	36.9	41,20	86,8	1,66	,631	4,51	13,60	1,40	810	988	47,3
14,40	625	36.0	47,20	85,2	1,73	,528	3,47	12,80	1,35	714	1180	47,6
12,40	880	36.4	38,30	92,7	1,50	,593	3,59	13,10	1,26	789	1260	46,3
12,40	740	34.7	39,40	96,7	1,69	,639	4,42	13,60	1,36	754	1160	47,0
12,20	616	38,3	43,70	104,0	1,39	,683	3,98	13,80	1,02	780	1090	48,6
11,50	592	40.5	44,30	107,0	1,71	,703	3,78	13,90	1,23	727	991	43,7
10,90	695	38,6	42,40	95,1	1,75	,726	4,26	13,00	1,77	667	894	40,2
12,30	835	37,5	41,00	101,0	1,44	,776	4,96	14,10	1,16	807	1190	44,5
10,10	655	39.4	42,90	95,5	1,62	,782	3,96	13,90	1,22	673	941	43,5
11,10	560	41.6	42,70	96,9	1,57	,784	3,69	13,80	1,13	748	805	46,0
14,10	780	37.4	38,60	85,4	1,67	,470	3,74	13,40	1,34	849	1170	47,4
10,30	664	35,8	41,50	92,9	1,64	,521	4,03	13,50	1,40	680	899	45,1
10,60	592	36.7	42,30	94,5	1,73	,696	3,88	13,60	1,13	675	1070	41,4
13,80	685	39,9	44,00	106,0	2,24	,555	4,33	14,80	1,38	734	842	50,4
14,00	759	40.1	42,80	96,3	2,39	,486	4,27	13,80	1,46	821	1100	45,3
10,70	1130	35.7	0,00	91,0	1,66	,851	4,88	14,50	2,21	786	0	44,9
12,30	1030	36,2	0,00	92,9	1,74	,662	4,71	15,30	2,12	858	0	46,7
11,50	605	38,8	43,60	0,0	1,74	,619	3,92	12,80	1,51	717	1120	55,3
5,16	319	36.4	26,40	57,9	1,49	,608	3,07	8,23	1,13	473	971	42,2
7,46	929	31.9	37,30	82,0	1,52	,602	3,47	12,30	1,17	416	1110	29,6
9,06	1070	38.0	38,20	90,6	1,74	,577	4,14	12,70	0,00	813	1380	47,0
11,20	811	39,9	43,00	99,6	1,85	,683	3,97	13,60	,49	848	1330	49,1
5,55	379	33.3	35,90	86,7	1,51	,811	3,75	11,20	1,46	541	1380	45,4
5,81	396	33,6	36,20	78,3	1,67	,580	4,01	11,10	1,68	161	1320	51,9
12,40	862	35,8	0,00	66,2	1,54	,768	3,79	14,00	1,61	858	0	47,7
8,74	360	25.3	52,10	114,0	1,45	,706	4,17	14,30	2,67	121	791	137,0
7,73	417	25.4	46,40	102,0	1,67	,632	4,52	16,70	1,72	126	743	19,6
8,56	518	28.9	49,60	112,0	1,52	,718	4,85	16,90	1,67	148	758	22,1
4.40	411	28.5	50,90	103,0	2,14	,572	5,09	14.40	0,00	164	1230	29,1
2,23	731	15,8	34,10	69,9	1,31	,475	3,37	9,57	1,07	116	126	7,3
2,71	636	15.3	38,20	79,4	1,46	,480	2,68	9,44	1,04	144	127	8,2
3,72	570	20.9	45,50	92,5	1,87	,461	3,82	12,10	1,57	172	199	9,1
3,44	404	19.7	43,20	91,3	1,79	,527	3,36	11,90	1,23	165	203	18,0
2,72	990	18.8	41,10	87,3	1,58	,399	3,51	11,00	1,38	148	212	12,0
4,63	955	21,3	54,10	93,1	1,79	,520	4,13	12,90	1,43	159	220	17,7
2,04	869	20.9	39,50	86,9	1,44	,486	9,07	9,39	1,93	148	495	18,5
2,24	853	25,6	35,80	79,3	1,66	,551	9,92	9,89	1,77	195	812	21,7
1,25	457	14.4	22,70	44,0	,78	,395	4,00	6,16	,84	88	285	13,5
3,47	598	25.3	36,50	76,9	1,48	,502	8,02	9,38	1,40	160	545	16,5
1,48	525	16,8	25,20	56,6	,99	,315	4,45	6,55	1,35	104	489	13,2
,85	186	12.9	17,20	36,6	,70	,222	2,95	4,04	,88	67	195	12,5
1,32	255	15.7	25,10	0,0	,90	,428	4,78	5,74	,90	113	513	17,2

Table VI. Limestone Hill Clays and

	SPECIMEN		K2C	FE2O3	NA2C	RB20
			PERCENT			
Yellow Limestone Hill Clays	B065	Beit Ummar	4,80	5,60	1060	113,0
	AR60	Arub	3,31	4,28	866	72,9
	JR65	Jerusalem	4,95	4,51	550	93,4
	JB60	El Jib	2,12	4,29	649	65,5
	JB70	W. of El Jib	5,65	5,83	907	125,0
Hesi, Azeka, Goded Pottery Matching the Limestone Clay Patterns	PB16		3,06	5,08	3060	74,1
	PB29		2,85	4,93	1890	67,2
	PC10		4,67	4,82	2800	71,8
	PE17		4,10	5,02	3160	61,5
	PF 4		3,19	4,16	3650	73,2
	PF 5		2,41	4,28	3290	42,8
	PF29		3,35	4,62	4810	64,7
	PG 2		3,68	4,65	1930	65,1
	PG31		3,87	4,61	3630	59,6
	GD 4	Tell Goded	2,60	4,47	1590	58,7
	GD 6		4,29	5,41	1620	114,0
	AZ 6	Tell Azeka	3,81	3,93	1580	72,1
	HS 7		3,52	3,61	1460	62,8
	HS 9		1,89	2,90	1660	27,7
	HS51		0,00	3,90	0	57,1

basis of ceramic style, six sherds were analyzed, plus a fragment of a base that was thought to be a mortorium (only rims have been reported). The base fragment, plus five of the mortoria, had identical patterns which fit well, with one exception, the general Mesopotamian pattern (Table VII). Close visual examination of the non-matching sherd, which had a red field clay type pattern, suggested that it did not belong with the others because of its paste differences. Mortoria of the matching element pattern were also obtained at Qeshet and Milha during the local tell survey. The disturbing exception to placing the mortoria in the Mesopotamian group is the Sc/Fe ratio (5.32 ± 0.39) which differs from the Mesopotamian ratio of 4.23 ± 0.14. A smaller group of mixed ceramic type and slightly different element patterns also fits into the Mesopotamian pattern except for the Sc/Fe ratio; this variation may be the result of heavy temper in these sherds. The temper has not been analyzed for Fe and Sc. Of course, it is possible that the true origin of these materials has not been found.

Ceramics with the Egyptian Nile clay pattern have not been found at this stage. Egyptian materials were expected because of the presence on the site of many small glass figurines considered in archaeological circles to be Egyptian imports. Attempts to match Hesi ceramics with Egyptian have been restricted to the Nile mud pattern.

The remaining pottery of the Hesi corpus fell mainly into groups of five to 12 sherds having slight variations with the general red field clay pattern (Table VIII). Attempts to pinpoint exact locations for these

Hesi Pottery of Matching Pattern

CONCENTRATIONS OF OXIDES

PARTS PER MILLION

CS2O	BAO	SC2O3	LA2O3	CEO2	EU2O3	LU2O3	HFO2	THO2	TA2O5	CR2O3	MNO	CCO
9,12	123	37.7	36,10	49,2	1,11	1,310	3,36	8,86	1,03	183	141	12,9
3,91	121	24.6	25,80	55,2	1,40	,473	3,97	6,90	,94	126	686	16,4
6,20	147	26.1	24,90	41,4	,98	1,140	2,58	6,22	,82	118	144	7,7
4,46	58	22.2	21,40	46,1	1,18	,319	2,49	5,82	,84	113	338	14,5
6,57	110	29.9	27,30	63,4	1,61	,434	2,91	7,72	,88	159	126	11,7
3,55	978	25.5	25,30	55,2	1,29	,364	4,06	8,63	1,45	122	322	19,4
3,69	304	24.4	22,90	52,5	1,44	,376	3,56	7,77	1,06	116	321	19,9
3,08	469	23.0	22,80	47,6	1,32	,317	2,66	8,67	0,00	128	371	17,2
3,94	413	28,7	28,40	65,5	1,72	,417	4,02	8,98	0,00	140	596	12,9
3,73	334	21.9	23,20	53,6	1,26	,452	3,06	6,79	,95	109	603	10,6
2,65	304	22,9	23,80	58,9	1,35	,437	2,93	6,96	,79	117	579	11,3
4,31	376	23.5	24,50	59,4	1,39	,451	4,79	7,45	1,01	0	478	11,3
2,72	468	19.8	20,10	44,5	1,30	,337	2,43	6,37	,66	105	269	13,4
3,08	92	20.9	19,60	49,1	1,27	,317	2,68	5,79	,50	99	281	14,8
3,57	673	24.0	28,40	57,0	1,46	,458	3,58	7,21	,98	128	396	13,0
5,48	692	28.5	31,60	62,9	1,59	,448	3,51	8,25	1,05	143	376	20,3
3,76	354	21.2	32,60	0,0	1,23	,251	3,45	6,42	,89	106	381	14,8
3,27	771	17.7	20,50	38,9	,98	,378	2,23	5,20	,55	90	303	12,4
1,71	582	15.0	18,70	30,9	,84	,344	2,17	4,17	,58	71	321	10,9
3,65	924	18.8	0,00	41,7	1,04	,455	2,48	5,54	,55	104	305	13,8

groups have not been made, and success in this area may depend upon analyses of more clay and pottery from a wider area than the initial survey.

There were a few sherds with totally unique compositional patterns, and these will probably not be placed, at least until much more is known of the elemental concentration patterns for the whole Near East. In this category is a group (1 in Table VIII) of nine sherds having the same pattern; unfortunately none of these have forms recognizable to the Hesi archaeologists.

Conclusion

Activation analysis was used to study 225 sherds selected from the excavated Persian level ceramics from Tell el-Hesi, Israel. The purpose of the project is to determine the origin of these materials—whether they were made at the site, came from sites local to the tell, or came from foreign sites. Establishment of ceramic trade patterns would help solve certain archaeological, historical, and biblical problems relating to this area in the time period considered.

Analysis of ceramics and clays from the coastal plain and hill country near Tell el-Hesi showed that while some of the hill country materials are distinct, the coastal plain pattern is similar to the point where separation of the materials by specific sites has not yet been done. Hesi materials do include sherds belonging to the hill country pattern, indicating trade relationships with sites in the Persian Province of Judah—an impor-

Table VII. Mesopotamian Samples

	Specimen		K_2O (percent)	Fe_2O_3 (percent)	Na_2O	Rb_2O
Mesopotamian Samples	NP 1	Nippur	.88	7.56	18200	39.7
	NM68	Nimrud Brick	1.70	6.25	11700	50.1
	TY 1	Tell Taya	2.15	7.26	7500	81.0
	TY 2	Tell Taya	2.67	6.58	9790	71.4
	TY 3	Tell Taya	1.16	5.57	6660	35.2
	AP 1	Arpachiyah	3.46	7.72	8240	119.0
	AP 2	Arpachiyah	2.48	8.83	5180	85.2
	AP 3	Arpachiyah	2.75	6.99	5500	85.8
	SM 1	Samarra Pots & Tile	.81	7.32	19600	32.6
	SM 2	Samarra Pots & Tile	.59	8.28	8440	21.5
	SM68	Samarra Pots & Tile	1.97	6.62	14400	62.2
	UR 1	UR Tablets	2.28	4.93	10200	65.3
	UR 2	UR Tablets	1.40	6.50	12100	57.7
	UR 3	UR Tablets	2.73	7.08	16500	55.0
	UR 4	UR Tablets	1.52	6.76	17700	60.5
	UR 5	UR Tablets	2.11	6.36	18700	68.3
	UR 6	UR Tablets	2.79	6.48	9770	68.6
	UR11	UR Pottery	2.51	6.05	22300	24.2
	UR12	UR Pottery	2.37	6.68	22800	67.8
	UR13	UR Pottery	1.51	7.57	25100	26.2
	UR14	UR Pottery	1.06	6.64	17500	28.5
	UR15	UR Pottery	2.11	7.20	23700	62.3
	CT 1	Ctesiphon	1.76	6.18	17500	73.5
	CT 2	Ctesiphon	2.41	6.17	17900	70.7
	CT 3	Ctesiphon	1.79	5.72	14600	75.8
	CT 4	Ctesiphon	2.03	6.99	11900	74.0
	CT 5	Ctesiphon	1.74	5.87	12300	62.9
	CT 6	Ctesiphon	1.32	6.69	6010	57.8
	NZ 1	Nuzi	2.37	5.95	7510	57.4
	NZ 2	Nuzi	2.20	4.83	5760	51.9
	NZ 3	Nuzi	2.25	6.08	5140	83.3
	NZ 4	Nuzi	2.24	6.02	4920	74.4
	NZ 5	Nuzi	2.22	5.93	6460	82.7
	NZ 6	Nuzi	2.02	6.04	4360	64.7
	BB 1	Babylon	1.14	7.42	19600	29.5
	BB 2	Babylon	1.71	7.19	20900	78.6
	BB 3	Babylon	1.91	7.35	19400	70.7
	BB 4	Babylon	1.78	7.33	20600	52.8
	BB 5	Babylon	1.81	7.00	18300	56.8
	BB 6	Babylon	1.94	6.83	19800	62.5
Hesi Sherds Matching the Mesopotamian Pattern	PF33		3.00	7.98	11800	49.9
	PB38		2.51	8.38	18900	70.4
	PE16		1.86	5.63	8110	64.5
	PA 6		2.62	8.29	13800	86.5
	PA 1		1.98	6.03	8890	67.8
Mortoria from Hesi and Nearby Tells with a Mesopotamian Pattern	GS 2	Tell Qeshet	2.13	5.67	7450	60.6
	ML23	Tell Milha	2.30	5.84	7690	60.8
	PG29		2.26	5.96	10100	48.2
	PF35		1.91	5.76	6480	57.3
	PF11		2.32	5.68	7170	52.4
	PC 6		1.83	5.84	7230	55.1
	PA26		2.32	6.34	8680	67.1
	PA29		2.30	6.23	7920	67.1

and Matching Hesi Sherds

CONCENTRATIONS OF OXIDES

PARTS PER MILLION

CS2O	BAO	SC2O3	LA2O3	CEO2	EU2O3	LU2O3	HFO2	THO2	TA2O5	CR2O3	MNO	COO
4,91	358	32.3	22,10	58,9	1,31	,461	3,77	8,40	1,24	495	1380	41,9
2,21	324	27.4	25,90	58,9	1,21	,383	3,94	6,74	1,10	659	1050	34,0
5,28	399	29.7	39,70	74,1	1,56	,429	4,71	10,30	1,39	414	1230	34,8
3,67	400	28.8	25,80	57,6	1,30	,539	3,84	7,46	1,08	844	1180	35,2
2,79	343	23.2	29,70	52,0	1,12	,332	3,48	6,44	1,02	623	951	32,1
7,68	561	31.1	26,80	72,3	1,28	,542	4,70	10,70	1,38	404	1120	33,1
4,68	235	38.5	29,00	55,0	1,17	,601	3,10	7,85	,95	550	1160	47,9
4,95	783	28.3	31,80	67,2	1,26	,451	4,07	8,88	1,23	531	1160	35,4
2,32	263	30.8	26,60	58,8	1,29	,448	3,22	8,21	1,51	363	1380	35,8
1,76	280	36.5	24,30	56,0	1,17	,465	3,96	8,44	1,56	407	1770	40,1
3,67	318	28.3	26,80	62,6	1,10	,373	2,89	7,97	1,19	308	1410	33,3
3,01	291	20.9	21,60	44,8	,80	,402	2,53	6,08	,88	226	742	19,9
2,79	359	27.4	23,90	48,6	1,23	,433	2,63	6,92	,90	362	1250	35,1
3,27	326	30.1	25,40	58,0	1,58	,385	2,99	6,77	1,22	445	1400	38,0
3,50	350	28.2	23,40	49,1	1,41	,368	2,74	7,25	,98	359	1190	36,1
3,87	271	27.3	21,30	56,7	,88	,358	3,38	7,09	,89	334	808	28,2
3,76	349	27.9	23,90	54,6	1,46	,364	3,01	7,05	1,66	382	824	27,6
1,64	598	26.1	28,20	60,1	1,41	,434	4,18	8,94	0,00	322	1170	24,0
2,58	345	29.4	26,90	60,6	1,42	,391	3,64	7,21	0,00	1040	1220	33,2
1,02	383	32.8	27,60	60,9	1,58	,430	4,09	8,73	,83	581	1210	37,5
2,89	220	28.4	24,20	57,0	1,01	,408	3,00	6,93	,62	406	1260	34,8
3,22	460	31.7	27,70	63,7	1,34	,427	3,64	8,07	,55	535	1170	34,3
2,88	408	26.3	24,90	56,0	1,12	,374	3,38	7,21	,98	492	1100	34,9
2,99	405	27.0	25,90	57,0	1,34	,411	4,11	7,18	,95	597	1040	27,8
3,18	366	24.8	24,20	55,0	1,25	,414	3,14	6,99	,89	454	1040	30,0
3,40	396	29.8	28,00	62,2	1,31	,415	3,32	7,66	,89	466	1110	36,0
3,00	393	25.4	26,00	55,8	1,30	,444	3,82	7,22	,95	432	953	25,4
3,20	595	24.7	41,70	78,1	1,82	,494	5,11	10,30	1,45	220	767	23,5
3,50	386	24.8	27,30	62,0	1,33	,410	3,92	8,06	,88	403	1030	25,3
2,29	341	20.6	21,50	48,7	1,28	,319	3,12	6,12	,83	379	901	25,7
3,93	858	26.1	26,50	57,2	1,23	,374	3,66	8,03	,92	355	1080	29,0
3,66	435	25.2	24,20	54,4	1,17	,453	3,29	7,69	,87	317	970	27,3
3,30	363	25.5	28,20	61,6	1,42	,450	3,92	7,92	,92	461	1010	27,0
3,55	373	25.7	26,40	57,2	1,19	,395	3,39	7,94	,87	360	958	26,2
2,18	357	31.6	28,10	62,2	1,34	,439	3,72	8,18	,91	540	1310	39,0
3,67	398	31.0	26,10	53,8	1,35	,423	3,48	7,78	1,02	402	1190	36,9
3,71	400	30.2	27,80	60,7	1,45	,443	3,32	7,34	,83	400	1290	38,7
2,75	417	32.7	25,80	57,1	1,56	,416	3,47	7,32	,94	601	1390	36,8
2,95	493	30.8	24,00	53,9	1,42	,411	3,19	6,67	,93	519	1210	37,2
3,12	0	30.6	25,60	52,4	1,28	,400	3,80	7,41	,85	431	1160	33,1
2,68	636	38.4	28,60	45,4	1,15	,558	2,68	5,92	,76	417	1510	33,1
2,94	485	41.5	25,30	49,7	1,55	,717	2,92	5,77	0,00	396	1650	40,0
3,17	310	28.7	23,90	51,2	1,20	,350	2,81	6,04	,83	422	1420	28,3
3,62	653	39.0	21,30	50,7	1,18	,454	3,32	6,09	,74	361	1290	34,4
2,99	504	28.5	27,90	61,7	1,38	,491	3,48	7,20	1,18	429	1260	27,8
2,59	918	31.2	27,60	47,6	1,24	,442	2,81	6,06	,96	392	1020	30,5
2,66	759	33.0	27,50	54,5	1,31	,545	3,04	6,46	,96	600	1050	27,3
2,96	277	32.5	25,30	57,4	1,44	,413	3,43	6,79	,83	536	1280	28,1
3,16	463	30.9	27,50	64,8	1,32	,440	3,37	7,07	1,26	875	1280	31,9
2,70	739	30.9	22,60	54,8	1,15	,398	2,67	5,76	,78	629	1380	25,1
2,46	629	33.5	23,50	52,6	1,26	,367	2,84	5,88	1,12	991	1110	27,4
2,98	646	34.5	26,80	58,8	1,26	,420	3,31	6,67	1,15	614	1450	30,8
3,31	431	31.6	28,60	61,5	1,35	,580	3,55	6,91	1,19	625	1250	29,1

tant historical and biblical fact. The largest group of Hesi ceramics fits the coastal plain pattern.

Clay and ceramic samples from the Aegean, Cyprus, Syria, and Mesopotamia areas were analyzed, and previous work on Egyptian materials was used as a basis of comparison with the Hesi materials whose elemental concentration patterns indicate foreign import. Sherds of Aegean/Cypriote origin were found. A small group of materials, including one stylistic collection of large bowls common in the Palestinian area,

Table VIII.

	SPECIMEN	K2O	FE2O3	NA2O	RB2O	CS2O	BAO
		PERCENT					
1	PB39	2.14	6,07	10300	81,5	1,31	851
	PC14	1.36	4,77	7110	37,7	1,40	1150
	PC15	2,20	5,79	9910	64,0	1,25	967
	PE12	1.80	5,59	8250	56,0	0.00	1320
	PE19	1.80	5,45	14300	0,0	0.00	872
	PE28	1,73	5,49	7550	57,7	1,17	1710
	PF42	1,85	5,12	10000	0,0	,86	2100
	PF43	1,71	5,32	8870	45,1	,96	1120
	PG11	1.80	4,22	6800	33,8	1,18	852
2	PG 4	1,94	5,57	8730	39,9	1,01	995
	PG 6	2,59	4,85	9860	33,2	,85	363
	PG12	2,24	4,43	9130	31,1	0.00	1180
	PG17	2,56	4,24	10100	31,7	,82	1050
	PG27	3,20	6,43	12200	48,8	1,84	1090
	PG30	1,62	5,54	8170	45,2	1,29	644
	PG33	1,73	4,76	9140	42,1	1,50	1150
	PG23	2,03	4,72	8500	46,3	1,31	368
	PB22	2,22	4,61	10200	42,9	1,05	2220
	PF19	1,57	5,61	8780	53,0	1,70	885
3	PA21	1,58	3,93	7140	31,8	,84	2130
	PB12	1,88	4,72	4370	43,3	2,38	453
	PF 9	1,10	4,73	4930	31,5	1,73	733
	PC 1	1,65	5,08	5170	51,1	2,22	626
	PC 7	1,75	5,33	5770	49,3	2,69	518
	PG13	2,07	5,19	5770	45,1	2,40	407
	PG39	1,91	5,22	5710	50,0	2,74	382
	PG35	,87	2,57	2910	19,4	,73	833
4	PA 4	1,06	4,42	5370	38,4	1,43	477
	PA18	1,15	4,30	4120	19,5	1,09	722
	PA23	1,27	4,03	5080	31,8	,95	1400
	PE10	1,50	4,88	3740	41,9	1,75	1960
	PF17	2,03	4,64	8850	38,8	1,06	687
	PG19	1,18	3,47	3750	25,3	,94	1020
	PG15	1,43	3,74	3860	27,7	1,24	405
	PG22	2,25	5,98	8830	42,0	1,15	903
	PG34	2,15	5,16	7860	30,0	1,15	1660
5	PB 8	2,38	5,47	10700	48,1	2,05	1860
	PG18	2,62	4,69	10800	40,6	,83	1460
	PE18	1,63	3,54	5390	26,6	,67	716
	PB10	1,33	3,29	4640	27,1	1,00	965
	PA10	1,31	3,88	0	37,7	1,42	913

appear to be from Mesopotamia. No objects of Egyptian Nile mud have yet been distinguished although such materials were expected from non-ceramic archaeological evidence.

To aid in grouping the materials, methods were worked out using ratios between the concentrations of pairs of elements and by computer-produced correlation plots between such pairs. Scandium and iron were highly correlated and of a distinct ratio for a given geographic area and clay type.

Miscellaneous Hesi

CONCENTRATIONS OF OXIDES

PARTS PER MILLION

SC2O3	LA2O3	CEO2	EU2O3	LU2O3	HFO2	THO2	TA2O5	CR2O3	MNO	CO
22.2	37.50	81.4	2.13	.667	10.40	8.08	.96	136	1170	23.6
19.6	41.60	73.3	1.93	.514	6.45	7.58	.66	181	973	17.9
22.0	38.70	78.1	2.28	.638	10.50	8.01	1.20	156	817	27.9
20.9	34.80	75.6	2.36	.559	8.56	8.54	.64	137	924	24.6
20.6	34.90	72.2	1.98	.567	10.50	7.53	0.00	130	987	21.2
20.1	34.30	82.0	2.08	.497	9.87	7.43	.94	153	939	22.5
19.7	35.10	70.4	1.89	.550	11.20	7.70	.82	152	945	19.9
19.5	33.30	71.3	1.90	.614	9.73	6.75	.66	140	888	20.1
17.8	37.60	74.6	1.65	.494	7.72	7.96	.86	153	829	16.6
20.6	33.70	74.3	2.43	.507	9.60	8.70	1.31	144	916	23.5
16.9	31.20	65.6	1.83	.424	7.94	7.69	1.21	115	713	17.0
18.0	29.90	70.4	1.95	.426	8.66	7.37	1.17	135	738	20.0
16.0	26.80	59.9	1.74	.420	8.86	6.57	1.50	120	592	17.0
24.0	37.00	77.7	2.64	.508	11.20	9.36	1.70	173	1020	25.3
21.0	34.20	74.5	2.28	.483	10.40	8.70	1.88	158	875	21.5
19.6	33.50	73.9	2.08	.445	9.79	8.29	1.29	133	848	18.4
16.9	30.00	64.7	1.55	.488	10.40	6.49	1.32	125	668	18.1
18.1	30.90	63.3	1.54	.464	14.40	7.64	1.55	131	657	21.7
20.9	36.80	65.0	1.71	.603	13.80	8.73	1.86	158	996	21.4
17.0	30.70	65.4	1.49	.431	9.26	7.30	1.70	138	508	16.8
21.5	36.30	76.9	1.68	.537	9.27	9.56	1.68	151	459	20.2
18.6	39.00	81.4	1.66	.545	10.60	8.88	1.89	166	310	21.4
21.7	38.30	82.6	1.82	.669	9.42	9.63	1.62	150	340	16.2
23.1	39.40	87.3	1.74	.629	10.10	9.90	2.17	178	483	19.6
23.1	40.80	95.2	1.96	.522	9.14	10.30	2.01	166	492	23.3
23.3	40.90	87.1	1.91	.699	9.59	9.93	2.18	152	498	20.9
9.9	15.50	39.0	.88	.289	4.30	4.54	.93	83	330	6.5
16.0	25.10	54.5	1.20	.364	5.38	6.48	1.12	183	430	16.0
16.9	31.20	72.8	1.33	.387	5.90	7.95	1.81	120	248	13.7
15.3	22.50	44.5	1.16	.301	5.56	5.75	1.19	93	392	13.2
19.4	33.70	72.1	1.51	.448	6.56	7.97	1.51	190	486	15.6
17.0	31.30	73.9	1.40	.425	8.54	9.24	1.39	132	588	19.4
12.9	28.10	62.4	1.21	.331	4.02	6.97	1.13	95	299	11.0
14.5	25.70	47.9	1.26	.398	5.61	6.01	.92	137	479	15.2
21.6	39.40	87.1	2.00	.515	7.49	9.06	1.52	181	641	19.4
19.1	34.70	75.6	1.80	.562	7.95	8.35	1.46	181	688	18.4
20.9	35.50	78.0	2.01	.523	10.10	8.67	1.74	144	841	20.0
17.9	33.00	71.0	1.88	.488	9.58	8.28	1.59	125	692	18.0
13.6	24.80	55.9	1.23	.323	8.07	6.31	1.09	114	386	13.9
12.7	25.80	53.7	1.18	.364	9.32	6.15	1.31	99	476	11.0
15.6	33.10	73.7	1.60	.570	10.50	8.70	1.64	139	577	15.6

Table VIII.

	SPECIMEN		K2O	FE2O3	NA2O	RB2O	CS2O	BAO
			PERCENT					
6	PF 1		1,76	7,20	11300	64,1	1,65	474
	PF23		1,61	6,39	10400	54,8	1,76	722
	PF27		2,76	5,42	11700	38,5	1,29	2620
	PF40		1,80	5,19	10000	50,7	1,81	644
	PG14		2,98	5,37	8890	39,8	,91	2690
	PG20		1,60	9,89	3480	73,5	2,93	499
	PG24		1,33	6,61	8420	41,7	1,71	774
	PE24		1,82	8,33	7560	77,3	2,01	482
	PB26		1,36	3,90	5400	32,6	1,37	2050
	PB31		1,28	6,01	9130	52,6	1,71	476
	PB37		1,41	6,36	3360	54,9	1,70	797
	PA35		1,46	5,52	7300	53,9	1,47	902
	PA30		1,71	8,51	7470	82,4	2,83	402
7	PB 5		1,64	6,74	5520	24,0	1,61	1690
	PB35		1,29	6,85	5270	20,0	1,53	2140
	PG36		,34	7,76	1540	28,4	3,03	182
	PG40		,60	7,43	2320	34,6	2,67	217
	PB 3		,34	7,56	1540	32,1	2,57	434
	PB 7		,48	7,54	1640	35,1	2,51	325
	PF30		,41	8,27	2240	34,0	3,02	285
	PA37		,69	7,62	2550	30,7	2,42	389
8	PA28		1,82	6,22	5000	67,1	2,96	730
	PG 3		2,30	7,29	6890	51,4	1,12	370
	PG10		1,48	8,56	2580	82,6	4,30	487
	PG26		1,81	8,70	6700	59,5	2,11	486
9	PF26		2,05	7,63	6750	67,0	3,07	616
	PB15		1,92	6,31	6160	63,0	2,64	1040
	PB20		1,20	5,75	7290	55,1	1,97	0
	PE13		2,42	5,16	11200	32,9	0,00	1210
	PB11		1,35	5,85	9080	49,2	1,58	850
	PB24		1,80	7,24	8260	63,9	2,30	508
	PB23		1,56	5,40	10300	44,3	1,52	552
10	PA 2		1,38	5,80	9660	53,4	1,52	0
	PA13		1,98	4,96	10400	50,9	1,50	1330
	PB13		1,35	4,67	7600	30,4	1,46	1270
	PF25		2,33	6,16	6490	35,7	1,08	657
	PG41		2,32	4,10	11000	30,5	1,05	542
11	NR 5	Neirob	1,89	4,90	4090	28,2	1,58	164
	HA 3	Hama	1,96	5,63	4040	35,5	1,47	338
	HA 4	Hama	1,18	4,26	3300	21,9	,79	509
	PB14		1,00	4,16	3880	28,3	1,48	729
	PA 3		1,67	5,84	6950	34,5	1,42	234

Acknowledgments

We thank the following persons and institutions for providing special materials and information: John Worrell, archaeological director of the Joint Expedition to Tell el-Hesi sponsored by the American Schools of Archaeological Research; Joe D. Seger, archaeological director of Hebrew Union College Biblical and Archaeological School in Jerusalem and di-

Continued

CONCENTRATIONS OF OXIDES

SC2O3	LA2O3	CEO2	EU2O3	LU2O3	HFO2	THO2	TA2O5	CR2O3	MNO	CO
		PARTS PER MILLION								
25.3	41,10	91,6	2,16	,624	7,86	8,89	2,15	145	1380	28,6
24.2	39,50	84,8	1,66	,550	10,10	12,00	1,87	158	1040	26,2
21.4	46,10	76,6	1,71	,540	10,30	10,50	1,85	150	823	22,0
19.0	33,20	71,7	1,74	,404	7,08	7,88	1,58	190	937	19,7
19.2	36,50	75,0	1,57	,473	8,08	8,24	1,95	145	972	21,8
38.8	65,60	139,0	3,63	,784	6,98	12,90	2,54	215	1150	40,2
24.4	39,00	84,2	2,03	,531	9,92	10,70	1,58	173	1050	27,2
31.9	53,30	119,0	2,70	,854	8,16	10,60	1,18	201	1570	43,2
16.1	25,70	61,9	1,11	,316	5,15	6,37	1,44	91	601	15,7
22.1	34,90	76,3	1,75	,471	9,13	8,68	1,92	153	936	23,9
24.3	45,50	87,9	2,13	,578	7,39	10,50	2,11	146	922	25,0
20.6	37,80	81,2	1,79	,510	7,60	8,40	1,69	150	880	20,4
32.3	51,40	107,0	2,71	,668	8,64	12,30	3,06	189	1160	47,3
24.5	49,70	118,0	2,19	,799	9,81	11,60	1,79	170	1240	37,9
24.0	39,90	114,0	2,28	,613	10,20	11,00	2,21	160	1340	35,8
28.9	74,60	162,0	2,85	,944	11,50	13,20	2,78	198	475	50,7
26.1	77,50	123,0	2,61	,838	9,77	11,90	2,64	175	1300	42,4
26.8	54,40	116,0	2,33	,854	9,44	12,00	1,93	202	1350	41,8
27.1	53,60	99,6	2,72	,708	10,20	12,80	2,74	199	1130	32,4
30.7	59,50	182,0	3,27	,784	12,50	14,00	2,62	233	1350	53,4
28.7	57,60	174,0	2,82	,689	11,90	12,80	2,49	214	1830	48,6
22.9	49,00	154,0	2,33	,779	14,90	12,20	2,33	188	2840	48,6
27.4	45,00	90,4	2,90	,617	0,00	9,11	2,26	161	1060	60,2
32.9	64,10	121,0	3,54	,771	9,15	14,00	2,07	179	1410	28,8
33.2	54,30	109,0	3,52	,693	8,92	12,10	2,22	189	1150	36,3
27.8	52,70	127,0	2,78	,781	17,30	13,40	2,62	224	1870	39,3
24.7	51,70	123,0	2,39	,660	15,10	13,00	2,37	219	2010	37,8
21.4	38,10	78,8	1,85	,550	9,96	9,84	1,96	158	1020	36,4
21.5	33,40	72,2	1,98	,513	9,02	7,46	1,52	140	797	30,1
21.8	36,30	78,2	1,74	,511	14,00	9,43	1,69	150	1060	31,7
27.1	45,80	97,4	2,09	,625	13,40	10,70	2,40	193	1410	36,7
20.7	36,00	74,0	1,61	,489	10,70	8,29	2,05	137	110	28,6
20.5	35,00	75,3	1,81	,641	10,40	8,10	1,59	162	681	20,8
18.2	34,60	74,2	1,73	,530	11,50	8,58	1,69	148	825	16,0
16.9	35,00	68,2	1,49	,460	6,88	8,06	1,73	120	899	11,4
22.7	43,50	94,1	1,90	,599	9,97	10,60	2,35	246	994	23,2
14.8	0,00	59,3	1,37	,401	8,62	7,54	1,49	116	857	16,9
15.6	23,40	48,1	1,25	,285	2,90	4,14	,97	164	520	19,7
18.6	29,30	59,9	1,50	,406	4,75	5,97	,99	319	749	24,9
13.8	21,90	44,3	1,10	,314	3,51	4,51	,96	185	465	18,2
14.4	23,70	49,2	1,14	,293	3,33	5,13	1,19	219	682	17,2
18.8	28,80	61,1	1,46	,419	3,05	6,06	1,32	146	706	19,6

rector of its excavations at Tell Gezer; Martin Aitken and Horace Catling of the Laboratory for Archaeology and the History of Art of the Ashmolean Museum, Oxford, who provided many of the Aegean materials; Barbara Clinkenbeard for the material from Lesbos; Ruth Amiran for the Nagila Excavation pottery; William Potts for the Gamma clay samples; Lotta Gaster for the Gaza samples and information; John Landgraf of the École Biblique et Archéologique Francais, Jerusalem for the

Beit Mirsim village materials and information. The Mesopotamian samples were provided mainly by the Metropolitan Museum of Art, New York City, and by the Peabody Museum, Cambridge, Mass. A few samples were from the Hartford Seminary Museum, Conn.

Literature Cited

1. Wright, G. E., *Biblical Archaeol.* (1971) **34,** 76-86.
2. Rainey, A. F., *Australian J. Biblical Archaeol.* (1969) **1,** 51-72.
3. Stager, L. E., *Biblical Archaeol.* (1971) **34,** 86.
4. Petrie, W. M. F., *Gerar* (1928) 8-9.
5. Wright, G. E., *Biblical Archaeol.* (1971) **34,** 83.
6. Worrell, J., *Israel Exploration J.* (1971) **21,** 232.
7. Asaro, F., Perlman, I., Dothan, M., *Archaeometry* (1971) **13,** 164-175.
8. Bullard, R., *Biblical Archaeol.* (1970) **33,** 98-132.
9. Perlman, I., Asaro, F., *Archaeometry* (1969) **11,** 21-25.
10. Abascal, R., Harbottle, G., Sayre, E., ADVAN. CHEM. SER. (1974) **138,** 81.
11. Bieber, A., Brooks, D., Harbottle, G., Sayre, E., *Intern. Conf. Appl. Nuclear Methods Field Works Art, Rome, May 1974.*

Bibliography

Amiran, R., "Ancient Pottery of the Holy Land," Masada Press, Jerusalem, 1969.

Lapp, P. W., "The Pottery of Palestine in the Persian Period" in "Archäologie und Alles Testament," A. Kuschke and E. Kutsch, Eds., pp. 179-197, J. C. B. Mohr, Tubingen, 1970.

Aharoni, Y., "The Land of the Bible," Westminster, Philadelphia, 1967.

Kenyon, K., "The Archaeology of the Holy Land," Praeger, London, 1964.

Wright, G. E., "Biblical Archaeology," Westminster, Philadelphia, 1957.

Crowfoot, J. M., Kenyon, K. M., Sukenik, E. L., "Samaria-Sebaste I-III," Palestine Exploration Fund, London, 1938, 1942, 1957.

Tufnel, O., "Lachish III," 2 vols., Oxford University Press, London, 1953.

Olmstead, A. T., "History of the Persian Empire," Chicago University Press, Chicago, 1970.

RECEIVED July 9, 1973. Research performed partly under the auspices of the U.S. Atomic Energy Commission.

6

Correlation between Terra Cotta Figurines and Pottery from the Valley of Mexico and Source Clays by Activation Analysis

R. ABASCAL-M.

Instituto Nacional de Antropologia e Historia, Mexico D.F.

G. HARBOTTLE and E. V. SAYRE

Brookhaven National Laboratory, Upton, N. Y. 11973

Activation analysis of selected groups of sherds of typically Oaxacan and Teotihuacan ceramics has shed some light on the question of the Oaxacan presence at Teotihuacan and the related concept of long-distance trade between the two centers during early Classic times. Studies of Preclassic "Tlatilco" figurines reveal them to have been locally manufactured.

The application of neutron activation analysis to the study of archaeological ceramics is now well established (*1*, *2*, *3*). Here we report our present method of operation, including techniques of data processing and taxonomy and some results obtained in studying archaeological material from the Precolumbian Valley of Mexico and the Oaxacan area. Under the rubric "Valley of Mexico" we discuss our work on two distinct archaeological problems: first, a study of Oaxacan-style pottery found in the Oaxacan barrio or suburb of Teotihuacan, and secondly, the probable origin of the well known preclassic terra cotta figurines unearthed in the burial ground of Tlatilco.

The first problem was posed to us by Evelyn C. Rattray, who also supplied the relevant sherds and clay samples. In one section of Teotihuacan, so many sherds of Oaxacan-style pottery have been found that archaeologists have come to regard the suburb or barrio as a possible site of Oaxacan occupation (*4*). Inasmuch as the home of Oaxacan pottery is located in the Monte Alban region some 400 km south of Teotihuacan, across mountains and desert, it is a question of considerable interest to learn whether the Oaxacan-style pieces found in Teotihuacan were manufactured in Oaxaca and transported by human carriers over this great distance, or whether they were simply produced in Teotihuacan out of local clays but in Oaxacan style. Clearly the archaeological implications of the two possibilities are quite different, involving questions of cultural contact and long-distance trade.

Figure 1. Teotihuacan provenance: typical Teotihuacan sherds

BNL No.	Rattray No.	Stratigraphy	Description
I-1	*Teo. 5*	*Surface*	*red on yellow basin (salt and pepper paste)*
I-2	*4868 N*	*layer 1*	*olla rim*
I-9	*5045 P*	*layer 3*	*comal*
I-10	*5045 R*	*layer 3*	*red-orange olla body*
I-3	*4944 T*	*layer 3*	*pol. bl. outcurving bowl*
I-4	*4946 N*	*layer 4*	*San Martin orange amphora*
I-6	*4958 Q*	*layer 5*	*SMO crater, pocked base*
I-7	*4981 M*	*layer 11*	*polished bl. br. outcurving*
I-8	*4990 S*	*layer 11*	*red rim basin*
I-37	*5023 O*	*layer 14*	*pol. br. bl. simple bowl*

The actual provenance of the Oaxacan-style sherds was a stratigraphic pit (designated N18W1) excavated by Rattray in a building (Site 7) located in grid square N1W6 on the Millon map of Teotihuacan (5). The period is within the classic Tlamimilolpan and Xolalpan phases, roughly 350 to 600 A.D. To provide a reference group representative of Teotihuacan a set of typical Teotihuacan-style sherds were taken from the same provenance. A group of five sherds of a Oaxaca-Teotihuacan hybrid style was also collected by Rattray from the same pit, N18W1. The local clay chosen was that used by the present-day San Sebastian

Teotihuacan potters, located in the southern part of the ancient city on grid square S2E1 of the Millon map. The genuine Oaxacan sherds taken for comparison were surface collections from two sites, Caballito Blanco and Dain Zu (Macuilxochitl), located 40 and 22 km (respectively) southeast of Oaxaca while the clay was from the source being used today by the famous potter Rosa of Coyotepec, about 12 km south of Oaxaca.

We also report on another group of sherds, of fine orange paste from Lambityeco (Oaxaca) supplied by John Paddock of the Instituto de Estudios Oaxaqueños, and finally a sherd of postclassic Mixteca–Puebla polychrome, supplied by Ignacio Bernal of the INAH (Mexico), of provenance Yagul, a site close to Caballito Blanco in Oaxaca (*6*). Photographs of all the sherds analyzed are shown in Figures 1 through 5.

Figure 2. Teotihuacan provenance: Oaxacan-type sherds

BNL No.	Rattray No.	Stratigraphy	Description
I-12	4944	*layer 3b*	*G 12, double line groove inside rim*
I-13	4946 H	*layer 4*	*G 21, parallel incised lines inside bottom*
I-14	4949 K	*layer 4b*	*G 1, apaxtle*
I-15	4957 J	*layer 4b*	*G 1, apaxtle*
I-16	4950 E	*layer 4c*	*— perforated sieve or censer*
I-17	4966	*layer 6*	*G 12, double line incising, inside rim, pol. gray*
I-18	4974 F	*layer 9-10*	*— dense gray, pol.*
I-19	4978 I	*layer 10*	*G 21, coarse incising, interior base*
I-20	5042	*layer 11*	*G 21, wave-like incising, interior base*

The second problem concerns the analysis of six terra cotta figurines from the site of Tlatilco. Tlatilco, a suburb of Mexico City, has been described by Piña Chan (*7*) in these terms: "De todos los pueblos preclásicos del valle de México que hasta ahora se conocen, Tlatilco fué el más cosmopolita, el más numeroso el más desarrollado cronológica y culturalmente." Although its importance has long been evident, Bernal (*8*) reminds us that "Tlatilco does not represent a city" but was rather "primarily a cemetery," though inhabited. There is no evidence of monumental sculpture or architecture; on the contrary its interest to archaeologists derives from the great wealth of small objects, ceramic fragments, jade ornaments, "yuguitos," and above all, figurines. These figurines, almost all representing women with short arms, heavy legs, and broad hips, are truly remarkable in their variety and animation. Many wear short skirts and some bear traces of red pigment, especially on the hair.

Figure 3. Teotihuacan provenance: Oaxaca-Teotihuacan hybrids

BNL No.	Rattray No.	Stratigraphy	Description
I-22	*4932 E*	*layer 1 (N19W2)*	*G 21, coarse incising on interior*
I-23	*4945 B*	*layer 3c*	*G 1, apaxtle*
I-24	*4950 D*	*layer 4c*	*— brown sieve, matte finish*
I-25	*4958 C*	*layer 5*	*— polished brown comal?*
I-26	*4981 A*	*layer 11*	*G 21, gray-brown, incised base*

Bernal (*6*) has described the extraordinarily close stylistic relationship of some of these figurines to those of the Olmec culture in the metropolitan Olmec zone along the Gulf Coast: "characteristic Olmec types are as fine and beutiful in Tlatilco as are their counterparts in La Venta. At times they are so similar that it would not be possible to tell their provenance had they not been found during explorations."

The question of these stylistic relationships and, more generally, of trade during Olmec (*i.e.*, 1300-400 B.C.) times is one of crucial importance to the correct formulation of the whole concept of an Olmec presence, empire, or influence in Mesoamerica. For example, Tolstoy and Paradis (*9*), in describing their most recent work in the Valley of Mexico, state that "this (Olmec) presence has been linked to trade by many (Coe, Flannery, Grove, and Jimenez Moreno), though the products ex-

Figure 4. Oaxacan provenance: matching group

BNL No.	Rattray No.	Stratigraphy	Description
I-32	*G 3*	*Caballitol Blanco*	*dense grey, polished grey slip*
I-33	*K 17*	*Dain Zu*	*brown matte basin, scraped exterior*
I-34	*G 12*	*Dain Zu*	*polished grey bowl, parallel incised lines*
I-35	*G 21*	*Dain Zu*	*matte grey, incised base*
F-121		*Lambityeco Md. 195 Cut S-4*	*barrel-shaped(?) vase, resist decor.*
F-122		*same, surface B*	*same, incised*
F-123		*same, N-11*	*polished both sides, red-brown outside, buff inside*
F-126		*Lambityeco S-8 Md.*	*Cajete, brown rim on orange*
F-129		*same, S-10 Md.*	*Rim, tecomate-like, incised line on ext. rim, highly polished outside, near-black zone around rim*
F-130		*Lambityeco bag 6060*	*rim, hemis. Cajete, type A-7 in Caso, Bernal, Acosta "La Ceramica de Monte Alban"*
F-131		*same, bag 6112*	*same*
F-132		*Lambityeco*	*flattened rim sherd, a fragment of object 6195(?)*
F-134		*Lambityeco*	*hemis. Cajete, incised-excised, traces of stucco obj. 7444*
F-135		*Lambityeco*	*frag. bat claw vessel, A-7 object 5914*

changed between particular communities cannot always be specified with any confidence." They discuss not only the possibilities of trade between the people of the Olmec presence in the Valley of Mexico and the Gulf Coast but of cross-tie interrelationships among the different preclassic sites within the Valley of Mexico—*i.e.*, Tlatilco, Tlapacoya,

Zacatenco, El Arbolillo, as evidenced by the material found in the stratigraphically corresponding layers in these sites.

It seemed to us that the whole problem of Olmec trade and the interrelationship of the preclassic sites in Mesoamerica offered an unusually favorable case for the method of grouping *via* paste-compositional patterns based on neutron activation analysis. As a start on this program we include here the analysis of six preclassic Tlatilco figurines, and we compare their analyses with those of two modern specimens and touch upon the archaeological conclusions possible.

Experimental

Analytical Procedures. The ceramic body is sampled as follows. First a fresh surface is prepared by removing with a tungsten–carbide burr a thin layer of the existing surface along with its dirt, slip, and in some cases weathered or eroded material. Then a tungsten–carbide drill is used to drill into the sherd along lines parallel to the surfaces, and a sample of about 100–200 mg is removed. With sherds too thin to drill, samples are removed from a fresh surface by further application of the tungsten–carbide burr. Clay specimens are not only analyzed as received but fine fractions are separated from them for additional comparative analysis. Samples (and standards) are oven-dried at 65°–70°C before use. Samples of 40 mg each are sealed into ultrahigh purity Suprasil T-20 fused quartz ampoules (U.S. Fused Quartz Co., Fairfield, N. J.) for irradiation in the nuclear reactor.

Figure 5. Non-matching sherds: descriptions given in Table III. Two modern sherds from San Sebastian Teotihuacan.

In all of our work we used as standards the six U.S. Geological Survey standard rocks designated AGV-1, BCR-1, DTS-1, PCC-1, GSP-1, and G-2 (*10, 11*). Weighed samples of all six, prepared and packaged in the same fashion as the ceramics, are included with each set of 20–30 samples and share the total bombardment history of that set. These standard rocks were analyzed by many laboratories for the many elements present in both substantial and trace amounts, and the results are reported by Flanagan (*10, 11*). Although Flanagan also tabulates the best values, we prepared our own table, eliminating excessively deviant values by applying Chauvenet's criterion to Flanagan's tabulation of original data. Our standard table is presented here as Table I.

Most of our neutron irradiations were done in the Brookhaven High Flux Beam reactor. Basically, two irradiations were made—a short one (0.5-3 min at 1×10^{14} neutrons/cm^2 sec) and a long one (typically 3.5 hr at 5×10^{14} neutrons/cm^2 sec). After the short bombardment, we counted short-lived radioisotopes of manganese, sodium, potassium, and lanthanum while the long bombardment served for the remainder (*see* Tables II–IV). In some instances the long bombardment was done first. Then, after 8–11 days decay, the same samples were reactivated by a short bombardment. They were then immediately counted for manganese alone (count times of 200 sec/sample suffice), and after it had died, a longer count (4000 sec) for all remaining isotopes was taken. Samples were counted with a 7% (*ca.* 35–40 ml) lithium-drifted germanium counter (Princeton Gammatech) having a resolution of 1.82 keV on cobalt-60, coupled to a 4096-channel Nuclear Data 2400 pulse height analyzer. The use of six distinctly different rock standards in each bombardment and the statistical comparison of the calibration constants derived for them constitute a rigorous test of the precision of the elemental determinations.

Data Handling. In previous studies where sufficient data were available to permit a proper statistical analysis of the nature of the distribution of the elemental concentrations in related groups of specimens, we have found our data to be logarithmically distributed. That is, plots of frequency of occurrence *vs.* log of concentrations closely approximated normal distribution curves. The use of log concentrations also allows one to give equal weight to a given fractional change, independent of the absolute value of the concentration range involved.

In more extensive studies we are now applying true multivariate statistical procedures, such as clustering and group discrimination based on generalized Mahalanobis distances (*12, 13*). Since, in the studies reported here, we had an insufficient number of specimens within each group to apply these procedures without significantly reducing the number of variates (elements) determined, we had recourse to simpler programs of data comparison and group testing. Because the compositional groups encountered were so clearly distinct, these simple procedures served well for classifying specimens.

One such program was the plotting of log concentration profiles for each sample through a computer program, POTPLOT, followed by simultaneous visual comparison of all elements in different specimens by superpositioning their plots over a light box. Such a procedure for specimen

Table I. Master

Oxide	Oxide Element Grav. Factor	U.S. Geological Survey Standard Rocks G-2 (97)	GSP-1 (38)	AGV-1 (64)
Na_2O	1.3479	4.067 ± 0.015 (35)	2.816 ± 0.017 (38)	4.255 ± 0.019 (35)
K_2O	1.2045	4.500 ± 0.018 (41)	5.507 ± 0.014 (41)	2.909 ± 0.082 (48)
Rb_2O	1.0936	189.5 ± 2.0 (12)	281.9 ± 5.2 (15)	76.2 ± 2.1 (15)
Cs_2O	1.0602	1.320 ± 0.045 (6)	1.100 ± 0.084 (5)	1.334 ± 0.041 (6)
BaO	1.1165	2108 ± 68 (18)	1515 ± 80 (19)	1418 ± 59 (18)
MnO	1.2912	354.1 ± 13.2 (54)	408.6 ± 11.2 (51)	965.0 ± 14.6 (54)
CoO	1.2715	6.01 ± 0.13 (15)	8.60 ± 0.29 (18)	19.38 ± 0.71 (21)
NiO	1.2726	9.26 ± 1.77 (16)	14.63 ± 1.76 (21)	23.15 ± 1.58 (21)
Sc_2O_3	1.5339	6.12 ± 0.34 (12)	12.19 ± 0.94 (11)	18.19 ± 0.49 (10)
Fe_2O_3	1.4297	2.696 ± 0.011 (41)	4.309 ± 0.016 (40)	6.808 ± 0.022 (43)
Eu_2O_3	1.1579	1.73 ± 0.14 (2)	3.18 ± 0.87 (2)	1.882 ± 0.087 (2)
Tb_2O_3	1.1510	0.60 ± — (1)	1.5 ± — (1)	0.89 ± — (1)
Dy_2O_3	1.1477	2.35 ± — (1)	6.50 ± — (1)	3.97 ± — (1)
Lu_2O_3	1.1372	0.20 ± — (1)	0.19 ± — (1)	0.42 ± — (1)
Cr_2O_3	1.4614	13.01 ± 0.70 (22)	19.14 ± 1.2 (20)	16.25 ± 0.79 (18)
CeO_2	1.2284	204.5 ± 9.8 (4)	553.8 ± 51.8 (5)	95.2 ± 14.0 (4)
ThO_2	1.1379	28.77 ± 0.31 (11)	125.6 ± 3.0 (14)	7.66 ± 0.23 (12)
HfO_2	1.1793	8.84 ± 0.98 (4)	14.56 ± 1.43 (4)	6.05 ± 0.72 (4)
Nb_2O_5	1.4305	21.4 ± 2.10 (9)	39.20 ± 1.09 (8)	27.84 ± 1.73 (5)
Sb_2O_5	1.3285	0.080 ± 0.004 (2)	3.61 ± 0.77 (4)	5.2 ± 0.84 (4)
Ta_2O_5	1.2211	1.11 ± 0.11 (2)	1.34 ± 0.39 (2)	
La_2O_3	1.1728	114.7 ± 5.2 (13)	297.2 ± 27.3 (16)	41.5 ± 4.8 (5)
Sb_2O_3	1.197	0.072 ± 0.004 (2)	3.25 ± 0.69 (4)	4.7 ± .76 (4)

[a] Values in ppm of oxide unless noted as %. Number of items in parentheses.

matching is roughly equivalent to the use of mean character difference clustering (*14*). Groups so formed were tested by another computer program, POTSTAT, which calculated and plotted the geometric mean of the concentrations with logarithmic standard deviation ranges. Specimens for which more than one elemental concentration was outside of 95% confidence limits for the group were eliminated.

In the earliest work on neutron activation analysis of archaeological material and in some later studies of other materials (*15, 16, 17, 18, 19*) it has been noted that the ratios of two elements in a group of samples can be less variable than the absolute amounts of either element. In another archeological study, one of the authors noticed that in certain Greek ceramics, nickel and chromium were correlated (*20*). Quite recently Bowman *et al.* (*21*) reported analyses of samples from an obsidian source in which a number of elements are highly correlated with iron (both negatively and positively); these authors noted that these orderly relationships can lead to positive identification of obsidian from this source. In such a case, the correlated elements varying over wide concentration ranges, any attempt to characterize this source or identify samples with it based on pure compositional clustering is bound to fail. The use of Mahalanobis distance or characteristic vector analysis would be appropriate (*13*).

Results of Analysis

The 29 pottery and two clay specimens from Teotihuacan, supplied by Rattray, were subdivided by her into four categories: (a) those with

List of Standards[a]

U.S. Geological Survey Standard Rocks

PCC-1 (16)	*DTS-1 (47)*	*BCR-1 (52)*
		3.232 ± 0.015 (33)
		1.689 ± 0.009 (47)
		53.8 ± 1.0 (13)
		1.280 ± 0.124 (6)
		840 ± 42 (20)
1172 ± 12 (40)	1227 ± 16 (47)	1816.2 ± 16 (51)
136 ± 5 (20)	167.6 ± 6.2 (20)	45.57 ± 1.06 (22)
3257 ± 132 (23)	3095 ± 104 (25)	16.87 ± 1.53 (20)
12.55 ± 0.91 (10)	5.89 ± 0.58 (7)	49.93 ± 0.89 (10)
8.326 ± 0.034 (41)	8.653 ± 0.028 (38)	13.472 ± 0.039 (44)
		2.461 ± 0.203 (2)
		1.2 ± — (1)
		7.17 ± — (1)
		0.68 ± — (1)
3909 ± 140 (18)	5885 ± 235 (21)	21.39 ± 1.24 (18)
		56.9 ± 4.7 (3)
		7.07 ± 0.16 (10)
		5.13 ± 0.53 (4)
		34.92 ± 6.78 (7)
	0.60 ± 0.04 (3)	0.84 ± 0.23 (4)
		1.07 ± 0.02 (2)
		33.0 ± 2.0 (8)
	0.54 ± 0.04 (3)	0.76 ± 0.21 (4)

typical Teotihuacan style; (b) those with typical Oaxacan style; (c) those with mixed Teotihuacan-Oaxacan stylistic elements; and (d) modern pottery known to have been made of Teotihuacan clay together with samples of the clay itself. Analysis showed that 10 of the 12 specimens with typical Teotihuacan style were very similar in composition, forming an acceptable compositional group in terms of all of the elements determined. Data for these specimens are given in Table II. The two exceptional specimens were quite different from the matching specimens not only in composition but also in microstructure; one was a very granular amphora, I-5, and the other was a thin orange bowl, I-11, with characteristic sizable mineral inclusions. The authors carried out an extensive analytical study of thin orange pottery.

Not surprisingly, all of the Teotihuacan-Oaxacan hybrid sherds had compositions very similar to the pure Teotihuacan style matching group (*see* Table II). Considerably surprising, however, was that nine of the 10 sherds with typical Oaxacan style also matched these two other groups in composition. Data for these specimens are also listed in Table II, with those for the non-matching sherd, I-21, which came from the lowest level, layer 14. This sherd however has a composition similar to most of those from Oaxaca (Table III). The analyses therefore indicate that most of these three stylistic groups of sherds found in the Oaxacan Barrio at Teotihuacan were distinctly similar in composition for all elements determined and hence must have been made from very similar clays. Most probably this clay would be the local clay of Teotihuacan. This probability was confirmed when it was observed that the two modern pottery specimens and the two clay specimens as they were received

Table II. Oxide Concentrations for Specimens

Specimen	No.	Percent Na_2O	Percent K_2O	Percent Fe_2O_3
Typical Teotihuacan sherds	I-1	2.41	1.68	5.95
	I-2	2.90	1.92	4.38
	I-3	2.36	1.57	4.76
	I-4	2.63	1.39	5.81
	I-6	2.77	1.52	5.10
	I-7	2.41	2.22	4.99
	I-8	3.02	2.18	6.15
	I-9	2.27	1.32	4.93
	I-10	2.85	1.49	5.95
	I-37	2.53	1.82	5.28
Oaxacan-style sherds	I-12	2.45	1.60	5.33
	I-13	2.20	1.30	4.94
	I-14	2.70	1.60	5.52
	I-15	2.60	0.93	5.91
	I-16	2.66	1.56	5.25
	I-17	2.27	1.15	4.84
	I-18	1.91	1.28	5.50
	I-19	2.61	1.79	5.60
	I-20	2.93	1.46	5.93
Oaxacan-Teotihuacan Hybrid sherds	I-22	2.96	1.30	6.07
	I-23	2.69	1.72	4.62
	I-24	2.33	2.10	5.82
	I-25	2.63	1.63	5.23
	I-26	3.34	1.31	4.85
Modern pottery	I-38	2.33	1.25	5.52
	I-39	2.20	1.28	6.52
Clays (as received)	C37	2.65	1.39	5.43
	C38	2.20	1.18	5.80
Fine fractions of above clays	C37F	1.14	0.94	6.29
	C38F	0.98	0.98	6.74

which are without question from Teotihuacan conformed to the same compositional pattern (*see* Table II).

From the correlation between the ancient pottery and modern clay one can draw several significant conclusions in addition to the most immediately obvious one that the matching ancient pots were locally made. The amount of temper in the ancient pottery, as evidenced by our comparison with the clay source, has not diluted or otherwise significantly altered the compositional pattern of the ceramic. Also the pottery must not have significantly changed in composition during burial. In addition we observed a small but significant difference between the composition

Table III. Oxide Concentrations for Rattray Specimens from the Matching

Specimen	No.	Percent Na_2O	Percent K_2O	Percent Fe_2O_3
Granular amphora from Teotihuacan	I-5	1.65	2.27	7.30
Thin orange incised bowl from Teotihuacan	I-11	0.22	4.31	5.07
Polished grey incised bowl of Oaxacan style and composition from Layer 14 at Teotihuacan	I-21	1.42	2.85	5.17
Red on pink, unslipped from Caballito Blanco	I-27	0.65	3.00	5.52
Polished grey with incised rim from Caballito Blanco	I-28	1.87	2.93	3.70
Coarse grey paste with cream and red-orange paint from Dain Zu	I-36	4.27	2.14	2.31
Polished grey dense Caballito Blanco	I-29	1.53	2.72	4.99
Apaxtle (thick) Caballito Blanco	I-30	1.64	2.97	5.02
Apaxtle (thin) Caballito Blanco	I-31	2.31	2.43	5.62

with Matching Compositions from Teotihuacan

						Parts per Million							
Rb_2O	Cs_2O	BaO	Sc_2O_3	La_2O_3	CeO_2	Eu_2O_3	Lu_2O_3	HfO_2	ThO_2	Ta_2O_5	Cr_2O_3	MnO	CoO
54.4	2.55	601	23.3	—	50.7	1.80	—	5.34	7.23	1.19	116	789	18.2
63.3	2.38	1280	17.5	—	44.7	1.61	—	5.04	7.17	0.74	102	601	15.0
66.0	3.50	1200	18.7	—	50.2	1.52	—	5.40	7.12	0.98	107	763	18.0
90.8	3.18	401	21.8	—	55.6	1.69	—	6.03	7.58	1.07	140	837	19.6
54.8	2.42	765	20.3	—	66.4	1.68	—	4.64	7.06	0.81	142	850	20.1
67.5	3.74	934	19.8	—	68.2	1.50	—	5.53	8.10	1.26	83	1030	18.3
59.5	2.44	985	24.9	—	64.8	1.96	—	5.78	6.95	1.03	139	892	22.8
52.4	2.68	1100	19.1	—	50.2	1.79	—	5.18	6.36	0.86	124	839	18.4
54.9	2.13	1090	19.8	—	41.4	1.46	—	4.56	5.90	0.81	138	675	15.9
67.8	2.45	673	21.6	30.5	60.7	1.91	0.334	5.14	8.10	0.76	108	802	17.5
59.4	3.63	613	20.3	32.9	62.7	2.02	0.550	6.11	7.20	1.06	85	1070	22.0
66.0	3.71	651	18.4	33.5	79.8	1.81	0.562	5.58	9.11	1.17	69	1050	21.3
54.3	2.52	919	19.3	30.0	52.3	1.69	0.483	4.79	7.31	0.94	114	686	16.7
71.8	3.29	350	20.5	28.3	52.8	1.76	0.505	6.08	7.22	1.16	125	737	18.2
63.5	2.78	861	19.0	26.4	49.2	1.60	0.438	5.17	7.44	1.31	140	779	18.7
80.0	4.40	680	18.6	36.5	72.2	1.99	0.682	6.51	9.77	1.53	72	1080	19.9
71.0	5.12	714	20.3	35.8	68.9	1.86	0.693	5.79	8.55	1.36	72	979	19.8
45.0	2.28	629	22.3	24.3	46.4	1.66	0.374	4.68	5.68	1.30	162	930	19.8
58.3	3.19	585	21.4	27.0	45.6	1.73	0.291	4.91	6.00	1.19	132	811	17.6
72.7	2.61	680	21.0	20.6	48.4	1.64	0.276	4.61	5.83	0.97	141	915	20.6
52.0	2.41	814	15.8	22.5	39.5	1.40	0.384	4.57	5.77	0.87	102	671	13.3
67.0	3.66	616	18.5	28.5	59.2	1.71	0.540	6.68	7.87	1.79	87	846	17.6
55.3	2.15	684	18.3	28.5	48.5	1.57	0.460	5.23	6.42	1.14	125	642	16.0
30.4	1.84	470	16.5	19.0	36.8	1.22	0.304	3.47	4.41	0.91	225	813	16.2
50.1	3.94	—	21.2	45.8	65.5	1.84	—	5.51	8.73	1.16	124	930	19.3
64.6	5.16	889	25.2	25.3	67.2	2.00	—	5.91	9.34	1.33	126	1290	21.9
64.8	4.16	597	21.7	—	62.9	1.94	—	5.51	8.73	1.24	135	1100	19.4
86.8	4.11	619	22.8	—	65.9	1.86	—	5.74	9.05	1.12	145	1040	19.6
104.0	4.29	604	26.5	46.9	86.7	2.40	0.548	6.28	10.90	1.00	93	1160	20.2
91.9	4.48	587	28.9	48.3	91.5	2.26	0.556	6.91	11.50	1.01	104	1190	19.3

of the clay as received and the fine fractions extracted from this clay. The primary difference is that the sodium concentrations in the fine fractions were in each case about half that in the total clay (*see* Table II). Since the ancient pottery coincides with the total clay rather than the fine fractions, it suggests that these ancient potters did not fractionate their clay.

An attempt has been made with Figure 6 to present graphically the overall matching data from Teotihuacan. The horizontal lines of the crosses indicate the mean oxide concentrations within the ancient matching sherds while the extent of the vertical lines indicates the spread of

from Teotihuacan and Oaxaca with Compositions Different Groups of their Find Sites

					Parts per Million							
Rb_2O	Cs_2O	BaO	Sc_2O_3	La_2O_3	CeO_2	Eu_2O_3	HfO_2	ThO_2	Ta_2O_5	Cr_2O_3	MnO	CoO
96.7	6.28	1040	37.6	—	42.4	1.34	3.42	4.86	0.60	522	1030	37.1
196.0	14.5	1060	27.4	—	122.0	2.30	6.56	16.0	1.51	116	673	19.2
102.0	8.66	1120	24.7	37.6	61.9	1.75	4.74	8.06	0.85	117	607	17.4
88.4	7.15	1590	29.7	52.7	114.0	2.47	5.79	8.93	1.00	110	554	17.6
99.3	23.8	1210	13.6	38.1	63.8	1.43	6.25	9.62	1.00	57	660	13.3
187	1.81	1840	5.2	21.8	55.0	2.91	1.71	0.75	0.23	17	353	10.4
107.0	19.30	1380	18.3	35.4	76.9	1.62	6.12	11.30	1.18	87	859	18.7
89.6	19.40	1550	18.9	38.0	78.2	1.63	7.01	11.90	1.19	91	895	19.1
87.4	13.90	1160	19.8	33.8	73.3	1.70	5.33	9.58	1.09	109	929	19.5

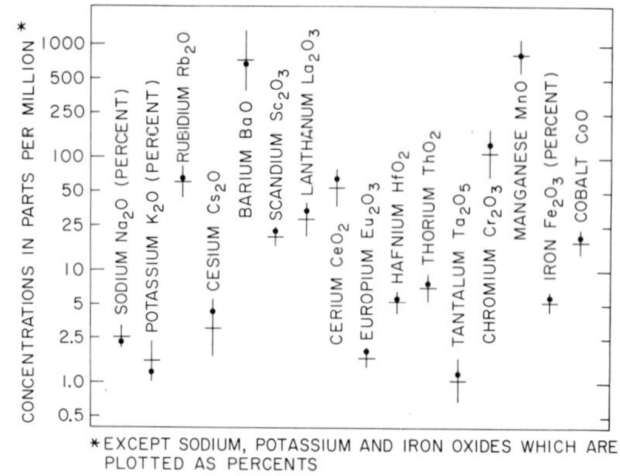

Figure 6. Comparison of the concentrations of many oxides in ancient pottery and in modern pottery and clays from Teotihuacan. Average concentrations (—) and 95% confidence ranges (|) of 23 ancient sherds. Average concentrations (●) in two modern pottery plus two clay specimens.

two standard deviation ranges—*i.e.*, 95% confidence limits for these data. Compared with these means and ranges are points representing the mean concentrations of the four modern specimens. Clearly the modern material is not significantly different in composition from the ancient.

From Oaxaca, Dr. Rattray supplied us with sherds of typical, local style pottery from Caballito Blanco and Dain Zu, together with a single clay specimen from Coyotepec. We were also fortunate in receiving for analysis other pottery sherds excavated in the Oaxacan area, including a group of 10 fine paste sherds from Lambityeco from John Paddock of the Institute of Oaxacan Studies; a more extensive study involving these

Table IV. Oxide Concentrations for Specimens

		Percent		
Specimen	*No.*	Na_2O	K_2O	Fe_2O_3
Rattray sherd from Caballito Blanco	I-32	0.90	2.53	6.12
Rattray sherds from Dain Zu	I-33	2.01	2.91	5.64
	I-34	1.43	2.75	6.08
	I-35	1.68	2.93	5.72
Paddock fine paste sherds	F121	1.34	2.86	6.38
	F122	1.12	2.75	5.65
	F123	1.15	2.33	5.96
	F126	1.14	2.61	5.79
	F129	1.34	3.30	7.49
	F130	0.89	2.42	5.61
	F131	1.04	3.17	5.50
	F132	0.93	2.58	5.58
	F134	1.04	2.13	5.27
	F135	0.88	2.72	5.40
Polychrome sherd	PC19	1.05	3.34	6.52
Coyotepec clay (as received and fine fraction)	C39	1.02	2.83	5.81
	C39F	0.64	2.52	6.36

sherds will be published separately. Another specimen was a Mixteca Postclassic polychrome sherd from Yagul which was provided by Ignacio Bernal of the Mexican National Institute of Anthropology and History.

Analysis showed that a number of these sherds can be arrayed in a group of approximately matching composition and that the single clay source also shows a reasonable compositional match to this group (Table IV). This should not, of course, be taken to indicate that we believe that in this clay source, Coyotepec, we have identified the place of manufacture of the compositionally matching group of Oaxacan pottery. To do this unequivocally it would be necessary to sample all the potential clay sources of the Oaxacan pottery and to determine that the variation from source to source was sufficiently great to eliminate ambiguity. Even then we could not be certain of identification since some clay sources used in antiquity may have been exhausted or otherwise disused. In this connection we plan to analyze the clay source reported by Paddock *et al.* (*22*) only 400 m from Lambityeco mound 190 and thought to have been worked in antiquity. With this source they could make pottery duplicating in appearance the Monte Alban IV style found in tomb 2 at Lambityeco. Clearly, more work needs to be done in surveying the clay sources of the Valley of Oaxaca and in analyzing the many types of Oaxacan ceramics. The presence of a number of non-matching sherds collected in the Valley of Oaxaca and analyzed by us (Table III) tends to confirm our belief that there were several clay sources in use and confirms Anna Shephard's opinion (*23*).

The placement of the Mixteca polychrome sherd from Yagul in the matching group of Table IV is of special interest. To the best of our knowledge this is the first piece of technical evidence that at least some of the famous Mixteca-Puebla polychrome was produced in the Valley of Oaxaca, confirming Bernal's view (*24*).

with Matching Compositions from Oaxaca

Parts per Million

Rb_2O	Cs_2O	BaO	Sc_2O_3	La_2O_3	CeO_2	Eu_2O_3	HfO_2	ThO_2	Ta_2O_5	Cr_2O_3	MnO	CoO
99.3	8.75	981	30.3	39.6	82.2	1.84	4.72	10.70	1.34	129	520	17.6
77.6	8.19	1110	24.1	43.3	80.3	1.94	5.62	8.94	0.96	84	1130	17.1
113.0	11.20	970	27.4	46.7	84.6	1.98	5.63	10.80	1.21	104	798	22.1
97.8	9.86	911	24.9	47.3	78.2	1.79	6.08	9.25	1.18	103	856	20.9
96.2	7.56	777	28.9	37.0	—	1.60	5.72	7.76	1.10	114	1210	21.2
113.0	6.93	751	26.9	39.0	—	1.48	5.76	11.10	1.49	114	467	13.5
99.9	6.12	709	24.1	47.0	—	1.64	6.15	10.60	1.50	117	460	15.6
99.5	6.11	833	23.6	40.0	—	1.75	5.66	10.20	1.40	110	541	14.8
140.0	8.77	819	32.1	50.0	—	1.94	5.42	11.20	1.46	131	601	21.5
116.0	7.11	856	27.8	36.0	—	1.62	5.16	10.20	1.21	104	525	15.6
113.0	6.97	709	27.4	46.0	74.5	1.76	4.86	9.32	0.98	98	683	14.7
103.0	7.45	—	28.9	40.0	88.0	1.71	5.12	10.90	1.11	113	553	15.3
102.0	5.65	—	24.0	39.0	68.8	1.72	5.28	10.50	1.31	113	430	14.5
135.0	6.30	—	26.6	37.0	66.4	1.67	5.14	10.40	0.89	104	481	14.4
128.0	7.40	766	29.1	39.0	—	1.91	7.50	10.80	1.13	109	953	15.5
129.0	6.68	867	28.4	—	79.5	1.93	5.77	10.90	0.77	142	651	17.3
147.0	7.43	648	34.6	51.6	93.0	2.02	4.59	11.60	—	126	369	11.6

Figure 7 compares the concentrations of the various oxides present in the Oaxaca (Coyotepec) clay and the average concentrations and spread of values observed in the matching Oaxacan pottery. The concentrations in the clay lie within 95% confidence limits for all elements but one. For the number of elements determined it is not statistically improbable that an individual specimen belonging to a group I element would lie outside such confidence limits, and we consider the compositional match between the clay and sherds to be good. This may merely mean that the Coyotepec clay is typical in composition of a range of clay sources throughout the Valley.

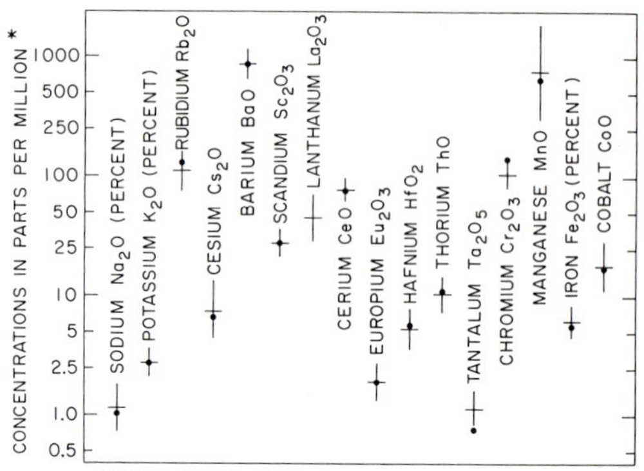

* EXCEPT SODIUM, POTASSIUM AND IRON OXIDES WHICH ARE PLOTTED AS PERCENTS

Figure 7. Comparison of the concentrations of many oxides in ancient pottery and in modern pottery and clays from Oaxaca. See *Figure 6 for legend.*

The Oaxacan composition differs from the Teotihuacan composition in many elements but particularly in the concentrations of the alkali metals. Figure 8 compares the mean concentrations and standard deviation ranges of the alkali oxides of some groups of these specimens with the average concentration of the Teotihuacan specimens which have either pure or hybrid Teotihuacan styles. Plotted values are the ratios of the concentrations in question to the mean concentration of this reference group. The centers of the diamond symbols, or crosses, mark the mean concentrations, and their vertical spreads indicate the spread of individual standard deviation ranges. The solid black diamonds are the ranges of the combined pure and hybrid Teotihuacan sherds, and the open diamonds are the same data for the Oaxacan style sherds of matching composition. Their compositional similarity is apparent. The crosses are the compositions of the compositionally consistent sherds from Oaxaca. Note that they are significantly different from the Teotihuacan sherds.

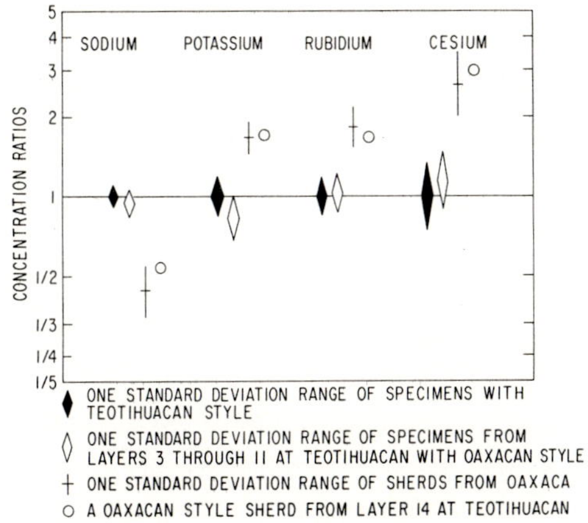

Figure 8. Ratios of concentrations in Teotihuacan and Oaxacan sherds to the average concentrations in sherds found at Teotihuacan with total or partial local stylistic characteristics

Figure 9. The figurines' descriptions and our laboratory numerical designations, reading from left to right, are as follows: figurine 3—height 7.7 cm, weight 21.9 grams, traces of red pigment on hair and generally over front of body; 2—height 8.0 cm, weight 23.8 grams, traces of red pigment; 5—height 11.4 cm, weight 42.8 grams, faint traces of red pigment overlying a cream-colored undercoat; 4—height 12.8 cm, weight 55.9 grams, red pigment on hair and skirt, underlayer of cream pigment over the entire body, face and skirt; 6—height 10.0 cm, weight 43.5 grams, red pigment overlying cream undercoat; 1—height 8.4 cm, weight 29.8 grams, trace of red pigment on headdress, fire-blackened in lower half. The clay is whitish gray.

Table V. Oxide Concentrations

Specimen	No.	Percent		
		Na_2O	K_2O	Fe_2O_3
Ancient figurines	TF-1	1.75	0.99	6.88
	TF-2	1.91	1.06	6.72
	TF-3	2.02	1.04	6.67
	TF-4	2.00	0.76	6.82
	TF-5	1.82	1.06	6.36
	TF-6	1.91	1.01	7.06
Clay from Tlatilco	TF-7	1.84	0.94	7.27
Leg of modern figurine	TF-8	2.21	1.10	6.28

The open circles show the relative positions of one Oaxacan style sherd from Teotihuacan which did not conform to the local Teotihuacan composition—specimen I-21, a polished grey incised bowl from layer 14, the lowest layer sampled at the Oaxacan Barrio. Our data are consistent with the possibility that this sherd was imported from Oaxaca. However, the preponderance of our findings suggests that the tendency was to import potters rather than pots from there to Teotihuacan.

Table VI. Comparison of Mean Compositions of Six Ancient Tlatilco Figurines with a Modern Figurine and Clay from Tlatilco[a]

Oxides Determined	Average Concentrations	
	Ancient Specimens	Modern Specimens
Iron (Fe_2O_3), %	6.75 ± 0.23	6.77 ± 0.70
Sodium (Na_2O), %	1.90 ± 0.10	2.02 ± 0.26
Potassium (K_2O), %	0.99 ± 0.11	1.02 ± 0.11
Manganese (MnO), ppm	810 ± 190	1040 ± 660
Barium (BaO), ppm	535 ± 26	515 ± 7
Chromium (Cr_2O_3), ppm	128 ± 6	158 ± 30
Cerium (CeO), ppm	68.9 ± 5.4	67.6 ± 7.6
Rubidium (Rb_2O), ppm	66.1 ± 4.0	55.3 ± 7.2
Lanthanum (La_2O_3), ppm	36.9 ± 2.3	35.1 ± 0.6
Cobalt (CoO), ppm	27.3 ± 1.8	27.1 ± 5.9
Scandium (Sc_2O_3), ppm	24.6 ± 0.6	24.8 ± 0.1
Thorium (ThO_2), ppm	8.99 ± 0.34	8.42 ± 0.01
Hafnium (HfO_2), ppm	6.37 ± 0.41	6.83 ± 0.17
Cesium (Cs_2O), ppm	4.60 ± 0.31	4.12 ± 0.25
Europium (Eu_2O_3), ppm	2.06 ± 0.17	2.14 ± 0.04
Tantalum (Ta_2O_5), ppm	1.06 ± 0.07	1.04 ± 0.11
Terbium (Tb_2O_3), ppm	0.91 ± 0.22	0.94 ± 0.04
Antimony (Sb_2O_3), ppm	0.53 ± 0.14	0.45 ± 0.16

[a] To simplify the presentation of means and standard deviations in Tables VI and VII, the standard deviations are presented as if the data were linearly distributed. Actually we believe our data to be logarithmically distributed and treat it accordingly. What are presented in Tables VI and VII are the geometric means of the groups of data with plus or minus one-half of the total standard deviation spread of the groups as calculated logarithmically. For reasonably closely matching groups the differences of notation are small and we felt that increased ease of reading justified the approximate notation.

for Tlatilco Specimens

Parts per Million

Rb_2O	Cs_2O	BaO	Sc_2O_3	La_2O_3	CeO_2	Eu_2O_3	HfO_2	ThO_2	Ta_2O_5	Cr_2O_3	MnO	CoO	Sb_2O_3
67.3	4.77	507	24.3	38.0	72.8	2.01	6.30	8.51	1.01	129	626	29.0	0.69
59.8	4.66	500	25.3	37.2	71.7	2.13	6.76	9.28	1.16	126	725	27.3	0.46
66.2	4.17	550	25.2	38.6	71.3	2.23	6.81	9.19	1.05	137	699	24.7	0.34
63.4	4.35	510	24.6	38.1	72.8	2.23	5.66	9.10	0.97	127	751	27.6	0.58
69.8	4.65	530	23.8	37.1	65.4	1.94	6.36	9.24	1.03	118	1120	29.4	0.45
70.2	5.05	550	24.9	32.4	59.4	1.80	6.31	8.60	1.11	133	964	25.8	0.67
50.2	4.30	520	24.9	34.7	73.0	2.11	6.95	8.41	1.11	179	1513	31.2	0.56
60.4	3.94	510	24.7	35.6	62.2	2.17	6.71	8.43	0.96	136	575	22.9	0.34

At this stage it is not possible to estimate how extensive the compositionally uniform clay deposits at Teotihuacan and Oaxaca are; they might be very extensive, and it is well known that the Becerra clay formation underlies large portions of the Valley of Mexico (*25*). One might even think that the entire Valley of Mexico contains only clay so uniform in composition that it would not be possible for one to differentiate pottery made from clays from different sites throughout it. This is not the case, however, as indicated by our second investigation—a study of the probable origin of the Tlatilco figurines.

The six figurines, all of Vaillant's class D-1 (*26*) and provenance San Luis Tlatilco, were supplied by the INAH (Mexico) and are shown in Figure 9. The modern specimens include clay dug at Tlatilco by one of us (R.A.-M.) and the leg of a fake figurine known to have been made in Tlatilco about 20 years ago (also shown in Figure 9). The analytical data for these specimens are given in Table V, and the mean concentrations and standard deviation ranges for the ancient figurines and modern specimens are compared in Table VI.

The results in Tables V and VI show that all six figurines have essentially the same compositional patterns and that the pattern for the figurines agrees exceedingly well with that of the modern samples. In obtaining these data we removed only small samples (50 mg) to avoid disfiguring the objects. We feel that some of the spread in the manganese data was partly the result of the samples' being too small to be truly representative. Manganese often tends to be distributed in a nonuniform particulate manner in pottery. Nevertheless, the impressive overall agreement argues quite strongly that the figurines were made of the local clay. We also analyzed several clays (kindly furnished by M. D. Coe) from the Olmec center at San Lorenzo Tenochtitlan; they were quite different from the Tlatilco clay and figurines in composition.

To return to our earlier point on the variation of clay compositions within the Valley of Mexico, Table VII shows the mean concentrations and standard deviation spreads of the Tlatilco, Teotihuacan, and Oaxacan matching groups. These data show that the Tlatilco and Teotihuacan clays can be as definitely resolved from each other as either can be from the Oaxacan.

Conclusions

We have shown by neutron activation analysis that groups of sherds from early classic layers at Teotihuacan, designated typical Teotihuacan style and hybrid, have very similar compositional patterns and presumably came from the same clay source. Analyses of modern pottery and clay from Teotihuacan again reveal a high degree of compositional similarity with the overall Teotihuacan sherd group, making it likely that the sherd groups above were locally made. Comparison of the Teotihuacan pattern with typical authentic Oaxacan sherds and clay shows significant differences while the Oaxacan clay and sherds conform to a consistent compositional pattern. Of 10 Oaxacan style sherds from Teotihuacan, nine had Teotihuacan compositional characteristics and hence were probably made locally by potters using Oaxacan stylistic traditions.

Table VII. Comparison of Concentration Means and Group Standard Deviations of Compositional Groups of Ancient Pottery Specimens

	Mean Concentrations		
Oxides Determined	*Tlatilco Figurines*	*Teotihuacan Sherds*	*Oaxacan Sherds*
Iron (Fe_2O_3), %	6.8 ± 0.2	5.3 ± 0.5	5.7 ± 0.6
Sodium (Na_2O), %	1.9 ± 0.1	2.6 ± 0.3	1.2 ± 0.4
Potassium (K_2O), %	1.0 ± 0.1	1.6 ± 0.3	2.7 ± 0.3
Manganese (MnO), ppm	800 ± 190	830 ± 140	680 ± 230
Barium (BaO), ppm	530 ± 26	730 ± 240	930 ± 220
Chromium (Cr_2O_3), ppm	128 ± 6	114 ± 33	107 ± 13
Cerium (CeO), ppm	69 ± 5	54 ± 11	77 ± 6
Rubidium (Rb_2O), ppm	66 ± 4	60 ± 13	105 ± 16
Lanthanum (La_2O_3), ppm	37 ± 2	28 ± 5	41 ± 5
Cobalt (CoO), ppm	27 ± 2	18 ± 2	17 ± 3
Scandium (Sc_2O_3), ppm	25 ± 1	20 ± 2	25 ± 4
Thorium (ThO_2), ppm	9.0 ± 0.3	7.0 ± 1.2	10.2 ± 1.1
Hafnium (HfO_2), ppm	6.4 ± 0.4	5.2 ± 0.7	5.6 ± 0.7
Cesium (Cs_2O), ppm	4.6 ± 0.3	2.9 ± 0.7	8.7 ± 3.2
Europium (Eu_2O_3), ppm	2.1 ± 0.2	1.7 ± 0.2	1.7 ± 0.1
Tantalum (Ta_2O_5), ppm	1.06 ± 0.07	1.07 ± 0.24	1.19 ± 0.18

However, one Oaxacan style sherd from Teotihuacan showed Oaxacan composition and thus was probably imported. Finally, the analysis of a group of six preclassic figurines and a clay from Tlatilco in the Valley of Mexico very strongly suggests that the figurines were also locally made.

Acknowledgments

We acknowledge the enthusiastic support shown by our archaeological colleagues, J. Paddock (Institute of Oaxacan Studies), E. Rattray (University of Rochester), and I. Bernal (INAH, Mexico), the wholehearted assistance of the Brookhaven Reactor Group, and of our technical assistants P. Donnelly and R. Graeser. We also thank G. Sanford for his work.

Literature Cited

1. Sayre, E. V., Chan, L.-H., Sabloff, J., in "Science and Archaeology," R. H. Brill, Ed., Chap. 12, M.I.T. Press, Cambridge, 1971.
2. Perlman, I., Asaro, F., *Archaeometry* (1969) **11**, 21.
3. Al-Kital, R. A., Chan, L.-H., Sayre, E. V., in "Hajar Bin Humeid: Investigation at a Pre-Islamic Site in South Arabia," Gus W. Van Beek *et al.*, Eds., The Johns Hopkins Press, Baltimore, 1969.
4. Milon, R., "The Teotihuacan Map," Vol. I, Part 1, p. 41, University of Texas Press, Austin and London, 1974.
5. Teotihuacan Mapping Project, *Science* (1970) **170**, 1077.
6. Bernal, I., in "Ancient Oaxaca," p. 345, J. Paddock, Ed., Stanford University Press, 1966.
7. Piña Chan, R., *Anales del I.N.A.H. (SEP)* Tomo IV 33-44 Mexico DF (1952).
8. Bernal, I., in "The Olmec World," Chap. 6, pp. 123-138, University of California Press, Berkeley and Los Angeles, 1969.
9. Tolstoy, P., Paradis, L., *Science* (1970) **167**, 344.
10. Flanagan, F. J., *Geochim. Cosmochim. Acta* (1967) **31**, 289.
11. Flanagan, F. J., *Geochim. Cosmochim. Acta* (1969) **33**, 81.
12. Mahalanobis, P. C., *Proc. Nat. Inst. Sci. India* (1936) **2**, 49.
13. Sneath, P. H. A., Sokal, R. R., "Numerical Taxonomy," Freeman and Co., San Francisco, 1973.
14. Sneath, P. H. A., Sokal, R. R., *cit. op.*, pp. 121-128.
15. Sayre, E. V., Dodson, R. W., *Amer. J. Archaeom.* (1957) **61**, 35.
16. Sayre, E. V., Murrenhoff, A., Weick, C. F., *Brookhaven Natl. Lab. Rept.* **BNL 508** (T-122) (April 1, 1950).
17. Stenger, V., *Anal. Chem.* (1971) **43**, 37A.
18. Rayudu, G. V. S., *et al.*, *Trans. Amer. Nucl. Soc.* (1968) **11**, 81.
19. Anders, O. U., *Anal. Chem.* (1972) **44**, 1930.
20. Harbottle, G., *Archaeometry* (1970) **12**, 23.
21. Bowman, H. R., Asaro, F., Perlman, I., *Archaeometry* (1973) **15**, 123.
22. Paddock, J., Mogor, J. R., Lind, M. D., *Bull. Oaxaca Studies* (Dec. 1968) No. 25.
23. Shepard, A., "Notas Sobre la ceramica de Monte Alban", In Caso, Bernal, and Acosta, "La Ceramica de Monte Alban" INAH-M 13.
24. Bernal, I., "Handbook of Middle American Indians," Vol. 3, Part 2, pp. 788 *et seq.* and 871 *et seq.*, University of Texas Press, Austin, 1965.
25. Coe, M. D., "*Mexico,*" p. 35, Ancient Peoples and Places Series No. 29, Thames and Hudson, London.
26. Vaillant, G. C., Excavations at Zacatenco, Anthropological Papers of the American Museum of Natural History XXXII-1 (1930); XXXV-2 (1935).

RECEIVED July 9, 1973. Research done under the auspices of the U.S. Atomic Energy Commission.

7

Neutron Activation Analytical Survey of Some Intact Medieval Glass Panels and Related Specimens

JACQUELINE S. OLIN

Conservation-Analytical Laboratory, Smithsonian Institution,
Washington, D. C. 20560

EDWARD V. SAYRE

Brookhaven National Laboratory, Upton, N. Y. 11973

The concentrations of 15 component oxides in medieval window glass were determined by instrumental thermal neutron activation analysis. Three groups of glass were studied: 52 specimens from a set of seven thirteenth-century French grisaille panels from a now demolished royal chateau at Rouen; 10 samples from a grisaille panel in the collection of the Princeton Museum; and a set of 32 random fragments of varied provenance. Significantly differing compositions were found. However the specimens from within individual and related groups of panels are compositionally similar even for different colors of glass. This similarity, therefore, indicates a common origin for the related pieces. Six of the random specimens had the same basic formulation as the specimens from the Rouen chateau panels.

The extent to which stained glass panels of the Middle Ages have been removed from their original settings in chapels and cathedrals, dismantled, and reconstructed with other glass to form new panels whose provenance must be established and the restoration pieces identified is often not appreciated. A window in the Princeton University Museum of Art is an excellent example of such reconstruction. The Princeton window depicts the martyrdom of Saint George (Figure 1). It was purchased for the art museum by the former Director, Frank Jewett Mather, Jr. in 1924. A similar window showing Saint George bound to a wheel was donated to Chartres Cathedral by either Guillaume or Robert de

Figure 1. Stained glass from Chartres Cathedral (in the Museum of Art, Princeton University). The top left quarter is from a thirteenth-century window depicting the martyrdom of St. George. Accession No. 71.

Coutenay during the second decade of the thirteenth century (*1*). The Chartres window remained there until 1788 when the sculptor Bridan petitioned its removal to allow more light to fall on his work on the new high altar. There is no direct record of what happened to the Chartres window, but modern grisailles now fill the space. Several studies of the Princeton window were published by W. Frederick Stohlman and by Henry Graham (*2, 3, 4, 5*); they concluded that only the top left quarter is original glass, and that it and a panel which remains at Chartres (Figure 2) as part of a composite of panels were once together at Chartres in the de Courtenay window. The possible location of two quarters of the de Courtenay window is still unknown. This is an example of the complexity which can exist in studying thirteenth century glass windows.

A large portion, possibly all, of the glass originally in a medieval window would probably have been produced by a single workshop. Of course, special colors and types of glass may have been brought in from other sources specializing in their production. However, it is unlikely that the same type of glass—*e.g.*, grisaille—in a single panel originally would have many sources. Glass produced by a single source of manu-

Figure 2. Thirteenth-century grisaille panels in the collection of The Cloisters Museum, New York. The panels came from the chateau of Rouen. Accession No. 69.236.2. Analytical data are in Table II. The numbers on the photograph correspond to sample numbers in the table.

Figure 3. Accession No. 69.236.3 (see *Figure 2*)

Figure 4. *Accession No. 69.236.4* (see *Figure 2)*

Figure 5. *Accession No. 69.236.6* (see *Figure 2)*

facture at a given time would probably have been consistent in its composition, and hopefully this composition would differ significantly from that of glass produced elsewhere. The uniformity of glass from a given source can be studied well by multiple sampling of individual panels because the likelihood of encountering glass of related origins within them is great even if later changes were made.

It is likely not only that glass of the same manufacturer will be found in a single panel but that separate fragments of individual glass preparations will be encountered. Such pieces should be very closely related in composition and hence could be separated through analysis. The extent and nature of such close correlation should indicate the size and uniformity of individual batches.

Figure 6. Accession No. 69.236.7 (see *Figure 2*)

This study was done to establish whether such correlations do exist and whether the distinctiveness of glass composition from different sources of glass manufacture will allow meaningful interpretation of observed correlations. A limited amount of historical information exists on the organization of the workshops which were the sources of stained glass in medieval Europe. In "A History of Technology" by Singer *et al.* reference is made to The Venerable Bede who in the eighth century stated in his "Historia Ablatum," that the French glaziers not only did the work required but taught the English how to do it. Frequent reference is made to the glazing of the church windows at Monkwearmouth,

Figure 7. *Accession No. 69.236.8* (see *Figure 2*)

Figure 8. *Accession No. 69.236.10* (see *Figure 2*)

England by French craftsmen who arrived in 675 A.D. Samples of window glass from the monastic site at Monkwearmouth have been excavated and analyzed by electron probe (6). Their sodium oxide concentrations range from 13.7 to 16.1%, and potassium oxide concentrations range from 0.24 to 1.12%. These alkalies are present in this glass in nearly the same proportions as in ancient Roman glass (7). In contrast, much of later medieval glass has been found by Geilmann and others (8) to have predominantly a potassium rather than a sodium composition. On the basis of this and other compositional considerations it has usually been concluded that wood ash was the main source of alkali in late medieval glass.

Figure 9. Thirteenth-century grisaille panel in the collection of the Museum of Art, Princeton University. Accession No. 43-65. Analytical data in Table V. The numbers on the photographs correspond to sample numbers in the table.

What we know of the techniques of glass making during the Middle Ages comes mainly from the description by Theophilus (9). According to him two parts of beechwood ashes and one part sand were mixed thoroughly and placed in a furnace to be fritted. The fritted mixture was distributed to white clay pots which were fired to melt the glass.

Analytical Procedure

Two procedures of sample activation were used in these analyses. In the first, samples of *ca.* 20 mg were weighed and heat-sealed in polyethylene tubing for a 20-sec activation at a flux of 1.5×10^{14} neutrons cm^{-2} sec^{-1} in the V11 facility of the high flux beam reactor at Brookhaven National Laboratory. ^{56}Mn was counted after about 4 hrs, and the ^{24}Na and ^{42}K were counted after 24 hrs. A second group of 50-mg samples was activated for 16 hrs at a flux of 1.5×10^{14} neutrons cm^{-2} sec^{-1}. After about two weeks the samples were poured from aluminum to glass vials for counting. The contents of the aluminum vials were washed into glass vials with ethyl alcohol for quantitative transfer. These samples were counted for 80–100 min for all other elements, using an automatic sample changer.

In the second procedure, one sample of *ca.* 40 mg was activated twice. The sample was first placed in a polyethylene vial and activated for 20 sec in the V11 facility at a flux of 1.5×10^{14} neutrons cm^{-2} sec^{-1}. After counting, the samples were transferred to quartz vials, reweighed, activated for 7 hrs at a flux of 2.8×10^{14} neutrons cm^{-2} sec^{-1} or 16 hrs at 1.5×10^{14} neutrons cm^{-2} sec^{-1}, and recounted in the quartz containers.

In both procedures, standard U.S. Geological Survey rocks G-2, GSP-1, AGV-1, BCR-1, PCC-1, and DTS-1 were weighed with each group of glass samples. Thus, glass samples and standard samples of similar weight were activated and counted simultaneously. Since these rock standards contained measurable amounts of all the elements determined in the glasses, they served as element-by-element monitors of the neutron flux densities and counting geometries encountered. For the rock standards we used the averages of all determinations reported after significantly deviant values had been eliminated by Chauvenet's criterion. For consistency we converted all elemental concentrations to oxide concentrations. The samples were counted using a 35-ml Ge(Li) detector; two 1600-channel SCIPP analyzers which could be connected in series to produce a 3200-channel spectra were used for the gamma-ray spectroscopy.

The data were collected on magnetic tape and fed into a CDC 6600 computer where a curve-fitting program, BRUTAL, was applied. BRUTAL output gave numerical values for the position and integrated intensity of each peak corrected for background. These intensities were punched on cards for each nuclide for each sample and standard. Using the CDC 6600 computer (in some cases the CDC 6400 computer of the Smithsonian Institution) and programs we have developed and named ELCALC and SAMPCALC, we calculate the oxide concentrations after correcting for radioisotope decay in both standards and samples. The specific activities of the U.S. Geological Survey rocks are calculated by ELCALC. They were used with the SAMPCALC program to compare the specific activities of each radioisotope in the samples and to calculate the oxide concentrations of the elements measured from these ratios.

Panels and Individual Pieces of Medieval Glass Studied

The first group we analyzed represents an extensive sampling—55 glass specimens from seven separate panels from the chateau of Rouen, built in the thirteenth century as a provincial residence for the French monarch. In the border of each panel appear castles of Castile, the well known insignia of Queen Blanche of Castile, mother of Louis IX. These panels are shown in Figures 2–8, in which the superimposed numbers indicate positions from which samples were removed. The chateau was demolished in the seventeenth century, and nine of its glass panels are now in the collection of The Cloisters Museum. We sampled seven of these panels, Accession Nos. 69.236.2, 3, 4, 5, 6, 7, 8, and 10, while they were disassembled for releading.

The second set consists of 11 samples from grisaille fragments in a panel which is part of the collection of the Princeton University Museum of Art (Accession No. 43-65) (Figure 9). Sampling was done when the panel was being releaded. At one time this panel was in the collection of a former curator of the Metropolitan Museum of Art, Bashford Dean.

Table I. Oxide Concentrations for Miscellaneous Identified Using Accession

	Percent				
Accession No.	Na_2O	K_2O	BaO	MnO	Fe_2O_3
23.229.2-2	0.76	12.4	0.111	0.79	0.53
23.229.2-3	0.30	22.3	0.492	1.83	0.35
23.229.2-10	1.29	16.7	0.181	1.01	0.81
23.229.2-14	0.78	16.0	0.082	0.36	0.91
23.229.4-1	0.21	20.8	0.182	1.01	0.49
23.229.4-6	0.61	12.1	0.090	0.61	0.69
23.229.4-7	1.73	12.0	0.230	1.09	0.59
23.229.4-10	0.68	12.0	0.052	1.26	0.41
23.229.5-2	5.66	4.5	0.186	1.10	0.83
23.229.5-5	19.40	2.4	0.042	0.44	1.19
23.229.5-6	1.95	16.6	0.082	0.74	0.58
23.229.5-7	2.02	18.1	0.220	1.93	0.48
23.229.5-8	3.96	10.6	0.184	1.77	0.52
23.229.5-9	0.88	11.7	0.036	0.74	0.63
23.229.5-10	13.20	2.4	0.241	0.65	0.72
30.73.210	2.83	11.4	0.123	0.37	0.82
30.73.211	15.00	4.8	0.046	0.51	1.00
30.73.212	3.04	12.6	0.179	1.25	0.46
30.73.214	1.25	12.7	0.167	0.97	0.54
30.73.216	0.63	10.7	0.060	0.40	1.09
30.73.217	2.44	11.0	0.124	0.65	0.84
30.73.218	1.15	12.7	0.190	1.13	0.69

[a] *See* Figure 10.
[b] ND—Not determined.

Dean has published photographs of this panel (*10*), and it has also been discussed by Paul Frankl (*11*). Its specific origin has not been firmly established, but all who studied it regard it as an excellent example of thirteenth-century grisaille glass.

The final group consists of 32 pieces of medieval glass from various sources which were part of the collection of The Cloisters Museum and were donated to the Corning Museum of Glass. Since the provenances of these specimens are not well established, the data for them cannot be used to characterize particular structures or glass workshops. These samples show the extensive variation in composition which does exist in medieval stained glass from different sources. Some, however, can be grouped upon the basis of similar composition.

Sampling during Restoration

The panels were all releaded and restored by Dieter Goldkuhle of Reston, Va. When the individual panes were separated, corners of those selected for analysis were cleaned with a tungsten carbide burr grinding

Non-Matching Group of Grisaille Fragments Numbers of The Cloisters Museum[a]

Parts per Million

Rb_2O	Cs_2O	Sc_2O_3	CeO_2	Eu_2O_3	HfO_2	ThO_2	Ta_2O_5	Cr_2O_3	CoO	Sb_2O_3
140	0.58	1.81	76	0.93	3.65	3.60	0.43	19.0	17.3	7.02
460	2.45	1.22	16	0.24	1.87	1.81	0.29	8.9	13.4	1.87
250	0.17	3.34	64	0.55	5.52	3.35	0.44	27.0	14.9	2.15
110	0.48	1.55	41	0.26	2.73	1.85	0.30	25.2	6.4	14.50
530	3.40	1.66	13	0.23	1.98	1.96	0.24	9.1	23.2	4.10
100	0.69	2.89	55	0.28	4.16	3.09	0.57	31.2	12.7	2.08
300	0.97	1.60	50	0.57	2.88	2.93	0.39	16.1	17.7	4.32
200	1.09	0.92	27	0.47	1.41	0.93	0.25	15.7	37.7	3.59
40	0.45	2.72	27	0.17	2.59	2.74	0.46	20.3	9.3	1.30
12	0.10	4.22	29	0.38	3.28	4.01	0.53	23.1	10.2	3.80
360	1.41	1.50	54	0.15	1.36	1.53	0.16	12.6	11.7	1.14
210	2.24	1.09	61	0.31	1.08	1.08	ND[b]	9.9	93.1	0.99
290	2.34	1.63	50	0.24	2.88	2.01	0.29	15.0	41.8	1.82
81	0.34	1.44	22	0.16	1.19	1.01	0.15	12.0	5.8	3.07
39	0.17	1.09	14	0.19	0.97	1.06	ND[b]	649.0	166.0	20.80
187	1.10	3.11	61	0.62	4.69	3.89	0.52	34.6	64.8	3.49
42	0.39	3.89	47	0.45	4.07	4.22	0.47	25.7	8.8	1.24
450	2.35	1.34	85	0.44	2.38	2.11	0.52	16.6	69.5	5.56
320	1.73	2.14	63	0.41	3.78	3.62	0.49	24.1	53.4	3.11
182	1.30	3.65	27	0.40	3.18	3.54	0.50	30.4	34.1	1.73
187	1.20	3.28	62	0.57	4.71	3.49	0.55	35.1	53.3	4.70
158	0.49	2.18	72	0.47	3.14	2.86	0.46	22.5	60.3	4.14

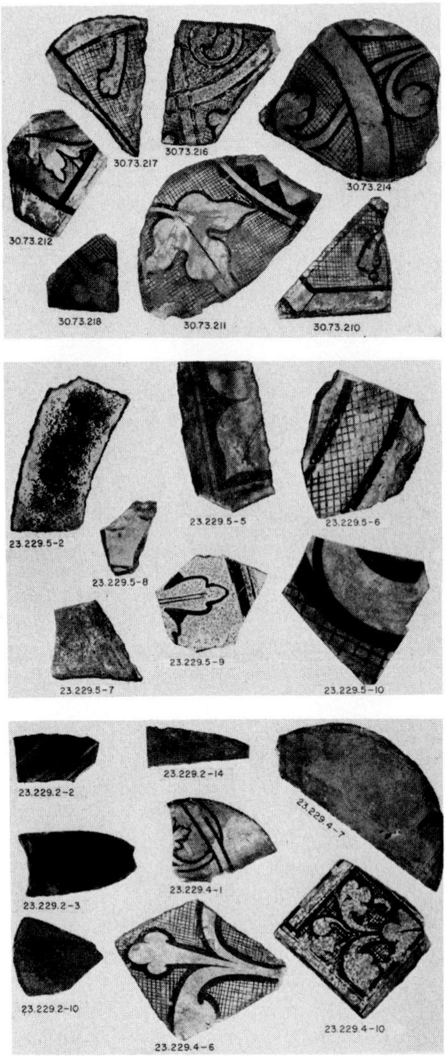

Figure 10. Pieces of grisaille glass from The Cloisters Museum. Analytical data are given in Table I.

Top: Accession numbers and attributions are 30.73.218: English, ca. *1290; 30.73.212: French or English, end XIII; 30.73.211: French or English,* ca. *1300; 30.73.210: French or English, early XIV; 30.73.217: French or English, XIII-XIV; 30.73.216: French or English,* ca. *1300; 30.73.214: French or English, end XIII-beginning XIV. Middle: Accession numbers and attributions are 23.229.5-2: French, XII; 23.229.5-5: Paris, XIII; 23.229.5-6: Bourges, XIII; 23.229.5-8: Rheims, XIV; 23.229.5-7: Paris, XIV; 23.229.5-9: Evereux, XIV; 23.229.5-10: Bourges, XIII. Bottom: Accession numbers and attributions are 23.229.2-2: striated red glass—Chartres, XII or XIII; 23.229.2-14: green glass—Paris, XIII; 23.229.4-7: Bourges, XIII; 23.229.2-3: glashed red glass—Paris, XIII; 23.229.4-1: Rheims, XIV; 23.229.2-10: Angers, XIII; 23.229.4-6: Bourges, XIII; 23.229.4-4: Bourges, XIII.*

tool to remove the surface hydrolyzed layer. The corners were scratched with a carbid-tipped stylus, and samples of about 100 mg each were broken off. This amount can be taken, and the fragment can be placed into the new leading without noticeable change to the panel. This opportunity to sample panels during restoration should not be overlooked because of the simplicity of the procedure at that time. Glass samples from the already separate glass pieces of the third group of specimens were removed in a similar manner.

Analytical Results

Medieval stained glass is generally understood to have been made with sand, wood ash, and perhaps lime. Because the wood ash concentration of potassium is high relative to that of sodium, one expects a high concentration of potassium oxide. All but four of the specimens analyzed confirmed this expectation, with potassium oxide concentrations between 11 and 32%. The exceptional four specimens had potassium oxide concentrations of 2–5% and sodium oxide concentrations of 5–20% and clearly were formulated with a different alkali. In fact, great variation was generally observed in the relative concentrations of alkalies in medieval glass, a situation that contrasts with the relatively few formulations encountered in ancient glass (7).

Table I lists analytical data on specimens of our set of random samples (Figure 10) which were selected to show the range of compositions which can characterize medieval window glass. Because of the marked differences in the concentrations of the constituents shown in this table, it is not likely that medieval glass of unrelated origins would be closely similar in composition.

In contrast to the diversity in compositions encountered in the unrelated specimens above, all 45 samples of uncolored or amber glass with grisaille painting from the Chateau of Rouen were basically similar in composition. This glass contains about 15% potassium oxide and only about 3.5% sodium oxide (*see* Table II). Because these similar specimens came from the same building, a common source of manufacture seems likely. These data therefore support our belief that individual sources of glass produced compositionally consistent products.

Consider the data on the colorless glass from the seven Rouen Chateau panels, panel by panel, as shown in Table II. Within the specimens taken from individual panels there are groups of fragments with exceptionally similar compositions. This is not surprising since a number of fragments probably were produced from a single manufacturing batch, and the fragments from the same batch would probably be used in the same panel. This seems to be the most logical explanation of this exceptional uniformity among specimens. In Table II, very closely matched specimens are grouped and printed in roman type. In one panel, 69.236.4, we encountered only specimens with very similar compositions. It would be interesting to analyze all the fragments of this panel to determine

Table II. Concentration of Oxides in

	Specimen No.	*Percent* Na_2O	K_2O	BaO	MnO	Fe_2O_3
		Ivy Panel 69.236.2				
Colorless	R013	4.0	13.0	0.122	1.12	0.59
	R014	3.9	13.8	0.128	1.12	0.61
	R011	*3.6*	*16.1*	*0.115*	*1.10*	*0.49*
	R012	*2.3*	*15.7*	*0.128*	*1.24*	*0.49*
Amber	*R017*	*3.9*	*15.1*	*0.116*	*1.01*	*0.49*
Blue	*R015*	*4.5*	*12.5*	*0.083*	*0.97*	*0.98*
	R016	*2.3*	*16.5*	*0.132*	*1.01*	*1.66*
		Ivy Panel 69.236.3				
Colorless	R019	3.2	14.9	0.116	1.05	0.49
	R023	3.5	15.5	0.103	1.13	0.49
	R020	4.1	16.1	0.126	1.07	0.58
	R022	4.1	16.0	0.136	1.05	0.57
	R018	4.1	14.6	0.125	1.16	0.57
	R021	4.0	12.6	0.118	1.02	0.57
	R024	4.4	14.5	0.114	1.20	0.56
		Ivy Panel 69.236.4				
Colorless	R028	4.2	15.8	0.124	1.17	0.56
	R029	4.1	15.2	0.110	1.09	0.55
	R030	4.0	17.7	0.115	1.07	0.55
	R031	3.9	13.9	0.117	1.08	0.55
		Ivy Panel 69.236.6				
Colorless	R01	4.1	13.4	0.122	1.12	0.59
	R02	4.3	14.1	0.119	1.24	0.61
	R03	4.5	14.3	0.118	1.14	0.61
	R04	*2.5*	*16.4*	*0.137*	*1.00*	*0.73*
	R05	*3.6*	*20.6*	*0.140*	*1.03*	*0.73*
	R06	*5.0*	*18.0*	*0.121*	*1.26*	*0.58*
Amber	*R09*	*4.4*	*13.8*	*0.133*	*1.09*	*0.56*
	R010	*2.8*	*17.0*	*0.131*	*1.02*	*0.65*
Blue	*R07*	*2.2*	*16.9*	*0.147*	*1.16*	*1.66*
	R08	*2.5*	*18.3*	*0.106*	*1.12*	*1.43*
		Ivy Panel 69.236.7				
Colorless	R034	3.2	13.5	0.113	1.06	0.49
	R037	3.5	12.6	0.107	1.11	0.49
	R032	*2.2*	*16.3*	*0.126*	*1.14*	*0.48*
	R033	*4.2*	*14.6*	*0.123*	*1.15*	*0.57*
	R035	*2.4*	*15.2*	*0.133*	*1.06*	*0.55*
	R036	*4.0*	*15.3*	*0.142*	*1.19*	*0.67*
	R038	*2.6*	*14.8*	*0.126*	*1.34*	*0.45*

whether they were formed from a single batch of glass. In general, however, we have encountered either more than one exceptionally closely matched group or other specimens which show a higher order of difference in composition.

Glass from the Rouen Chateau Panels

Parts per Million

Rb_2O	Cs_2O	Sc_2O_3	CeO_2	Eu_2O_3	HfO_2	ThO_2	Cr_2O_3	CoO	Sb_2O_3
				Ivy Panel 69.236.2					
322	1.64	2.08	51	0.55	2.38	1.90	20.0	28.3	4.2
344	1.36	2.22	61	0.50	2.70	2.05	18.0	30.0	4.2
288	*1.37*	*1.74*	*54*	*0.48*	*2.77*	*1.89*	*15.6*	*20.0*	*3.5*
225	*0.56*	*1.73*	*44*	*0.47*	*3.60*	*2.13*	*16.6*	*20.2*	*4.0*
252	*0.89*	*1.81*	*83*	*0.38*	*1.86*	*1.44*	*16.8*	*17.0*	*0.4*
255	*1.05*	*1.86*	*41*	*0.48*	*1.88*	*3.96*	*27.0*	*870.0*	*39.4*
141	*0.92*	*2.10*	*78*	*0.41*	*2.79*	*5.65*	*30.5*	*1390.0*	*237.0*
				Ivy Panel 69.236.3					
331	1.03	1.74	54	0.41	2.31	2.06	13.4	17.3	2.9
336	1.15	1.70	55	0.35	2.33	1.67	15.3	20.9	2.3
388	1.16	2.25	63	0.62	2.55	2.13	17.9	25.2	3.3
383	1.18	2.12	60	0.59	2.45	2.16	19.2	24.0	3.1
370	1.28	2.12	61	0.58	2.76	2.07	18.2	26.0	4.2
384	1.17	2.20	61	0.52	2.45	1.92	17.1	24.4	2.9
341	0.82	2.10	55	0.59	2.73	1.91	15.3	25.8	4.3
				Ivy Panel 69.236.4					
365	1.24	2.10	58	0.50	2.69	2.32	13.3	28.3	4.0
345	1.01	2.02	52	0.49	2.40	1.94	15.5	26.6	3.9
353	1.06	2.09	57	0.51	2.69	1.98	15.4	27.8	3.7
345	1.29	2.13	55	0.56	2.65	2.10	20.3	25.6	4.0
				Ivy Panel 69.236.6					
322	1.20	2.12	56	0.58	2.84	1.83	18.6	26.2	4.2
329	1.75	2.75	53	0.56	2.76	2.06	19.2	29.3	4.5
340	1.28	2.26	59	0.61	2.45	1.91	18.3	24.7	3.5
198	*0.75*	*2.31*	*87*	*0.62*	*3.49*	*2.64*	*20.0*	*25.7*	*3.7*
181	*0.52*	*2.15*	*79*	*0.58*	*3.53*	*2.53*	*18.9*	*45.9*	*15.5*
348	*1.67*	*2.11*	*58*	*0.48*	*2.58*	*1.75*	*15.1*	*28.6*	*3.5*
210	*0.68*	*1.96*	*73*	*0.48*	*2.84*	*1.95*	*14.9*	*24.1*	*3.2*
173	*0.62*	*1.90*	*77*	*0.48*	*3.15*	*2.07*	*15.7*	*16.8*	*2.4*
175	*0.48*	*2.25*	*84*	*0.65*	*3.43*	*5.59*	*22.5*	*1220.0*	*177.0*
75	—	*1.09*	*49*	*0.51*	*1.94*	*4.13*	*16.1*	*890.0*	*132.0*
				Ivy Panel 69.236.7					
346	1.26	1.69	34	0.49	2.05	1.57	13.3	16.6	3.1
333	1.12	1.69	45	0.48	1.95	1.67	16.0	18.9	4.6
228	*0.55*	*1.62*	*35*	*0.34*	*3.68*	*1.96*	*14.8*	*19.5*	*2.9*
359	*1.08*	*2.16*	*45*	*0.61*	*2.55*	*1.77*	*17.6*	*25.5*	*4.5*
221	*0.55*	*1.97*	*53*	*0.46*	*2.40*	*2.28*	*14.2*	*25.5*	*0.3*
207	*0.75*	*2.04*	*84*	*0.58*	*3.17*	*2.22*	*16.7*	*21.3*	*1.4*
243	*0.53*	*1.64*	*40*	*0.39*	*1.86*	*2.75*	*12.6*	*19.1*	*5.8*

Table II.

	Specimen No.	Percent Na_2O	K_2O	BaO	MnO	Fe_2O_3
		Ivy Panel 69.236.8				
Colorless	R039	2.5	17.1	0.120	1.27	0.46
	R043	2.5	15.0	0.124	1.25	0.45
	R041	3.1	14.1	0.101	1.06	0.51
	R045	3.3	15.8	0.111	1.10	0.48
	R040	*2.2*	*17.2*	*0.129*	*1.15*	*0.57*
	R042	*3.9*	*13.7*	*0.121*	*1.09*	*0.55*
	R044	*3.1*	*14.9*	*0.094*	*1.07*	*0.45*
Blue	*R046*	*2.1*	*16.9*	*0.145*	*1.08*	*1.71*
		Acanthus Panel 69.236.10				
Colorless	R049	2.7	15.2	0.144	1.31	0.58
	R050	2.7	16.1	0.147	1.36	0.49
	R051	2.6	16.7	0.142	1.32	0.49
	R055	2.7	14.8	0.142	1.30	0.58
	R056	2.7	16.3	0.149	1.36	0.53
Red flashed	*R048*	*2.5*	*16.2*	*0.113*	*0.98*	*0.48*
Amber	*R052*	*2.4*	*15.8*	*0.107*	*0.84*	*0.41*
Green	*R053*	*3.0*	*15.8*	*0.122*	*1.00*	*0.60*
Blue	*R054*	*1.9*	*10.7*	*0.117*	*0.85*	*0.96*

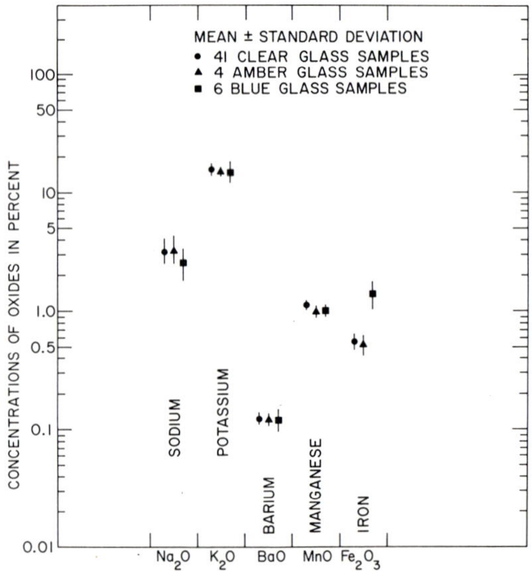

Figure 11. Comparison of mean concentrations of major and minor components in colorless, amber, and blue glass in the grisaille panels from the Chateau of Rouen

Continued

Parts per Million									
Rb_2O	Cs_2O	Sc_2O_3	CeO_2	Eu_2O_3	HfO_2	ThO_2	Cr_2O_3	CoO	Sb_2O_3
				Ivy Panel 69.236.8					
240	0.59	1.67	43	0.34	3.39	1.70	14.7	19.4	6.3
241	0.59	1.61	39	0.45	3.18	1.73	13.1	18.7	4.4
329	1.43	1.67	48	0.58	1.86	1.41	18.0	17.6	5.9
330	1.22	1.64	42	0.41	2.02	1.63	16.8	16.9	2.3
252	*0.82*	*1.99*	*65*	*0.39*	*3.15*	*2.48*	*19.8*	*20.9*	*1.4*
300	*1.60*	*1.97*	*49*	*0.49*	*2.35*	*1.75*	*14.3*	*26.4*	*3.0*
316	*1.21*	*1.56*	*33*	*0.46*	*1.85*	*1.44*	*12.3*	*14.5*	*1.6*
227	*0.48*	*2.12*	*96*	*0.76*	*3.39*	*4.02*	*20.4*	*1170.0*	*130.0*
				Acanthus Panel 69.236.10					
275	0.92	1.81	92	0.51	2.69	2.15	14.0	42.0	1.2
284	0.70	1.75	91	0.50	2.46	2.14	10.0	42.1	1.7
292	1.03	1.81	96	0.60	2.39	2.19	14.3	39.8	0.8
359	0.87	1.94	103	0.56	2.66	2.14	15.2	45.5	1.8
326	1.03	1.66	96	0.40	2.38	2.15	15.2	40.4	1.9
232	*0.75*	*1.54*	*84*	*0.43*	*2.38*	*2.03*	*20.9*	*23.0*	*4.4*
266	*0.83*	*1.49*	*75*	*0.37*	*2.61*	*2.11*	*14.7*	*14.0*	*0.3*
279	*1.14*	*1.98*	*80*	*0.38*	*3.41*	*2.54*	*11.5*	*59.4*	*113.0*
264	*0.47*	*1.77*	*92*	*0.50*	*1.59*	*4.76*	*19.4*	*1300.0*	*43.7*

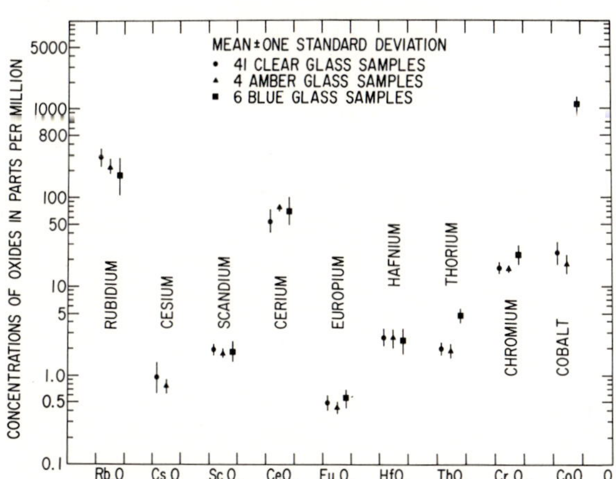

Figure 12. Comparison of mean concentrations of trace components in colorless, amber, and blue glass in the grisaille panels from the Chateau of Rouen

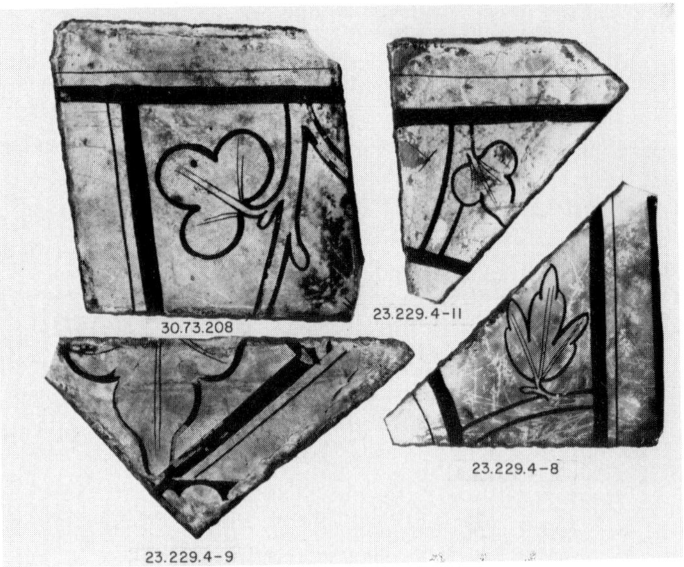

Figure 13. Pieces of grisaille glass from The Cloisters Museum. Accession numbers and attributions are 23.229.4-8: Paris, XIV; 23.229.4-9: Chartres ?; 23.229.4-11: Chartres, ? XIV; 30.73.208: French, ? XIII-XIV. Analytical data in Table III.

In comparing glass from different panels one observed in general only about the same order of difference that exists between different fragments in the same panel which do not correlate very closely. Most of those differences probably represent the normal variation between batches of glass from the same workshop.

A comparison among the average composition of the colorless, the amber, and some of the blue glasses from the Rouen Chateau is shown in Figures 11 and 12. For most elements the standard deviation ranges of concentrations encountered in all three colors overlap, and hence the average concentrations in each of the colors were not significantly dif-

Table III. First Matching Group of Grisaille Fragments Identified

	Percent				
Accession No.	Na_2O	K_2O	BaO	MnO	Fe_2O_3
23.229.4-8	0.33	18.2	2.54	1.21	0.44
23.229.4-9	0.30	18.2	2.19	1.16	0.31
23.229.4-11	0.30	19.4	1.42	1.22	0.41
30.73.208	0.30	17.9	2.69	1.32	0.43
Mean	0.31	18.4	2.15	1.23	0.39
Group std dev., %	4.9	3.6	33.5	5.5	17.6

ferent from each other. However, in the blue glasses the cobalt, iron, thorium, and chromium concentrations were all higher, but this could be a result of a cobalt-containing material added as a colorant. In all the colored glasses, and in the green and flashed red glasses analyzed, the same basic glass seems to have been used. Therefore, all of the above glass was probably produced by a single workshop.

Three additional pieces of blue glass were identified as restoration glass on the basis of their physical appearance before sampling. Samples R025, R026, and R027 were taken from these pieces which were all in panel 69.236.4 shown in Figure 4. Samples R025 and R026 are not shown because the entire top area of that panel was replaced during recent restoration. Figure 4 shows the recently restored panel. The concentra-

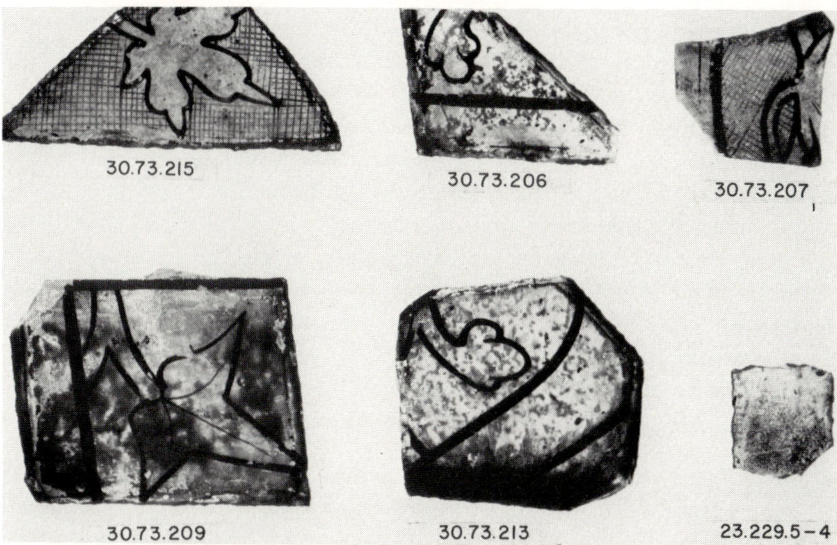

Figure 11. Grisaille glass from The Cloisters Museum. The accession numbers and attributions are 23.229.5-4: Chartres, XIV; 30.73.206: French or English, late XIII; 30.73.207: French or English, early XIV; 30.73.209: French or English, beginning XIV; 30.73.213: French, ? XIII-XIV; 30.73.215: French or English, XIV. Analytical data in Table IV.

Using Accession Numbers of The Cloisters Museum (*See* Figure 1)

Parts per Million

Rb_2O	Cs_2O	Sc_2O_3	CeO_2	Eu_2O_3	HfO_2	ThO_2	Ta_2O_5	Cr_2O_3	CoO	Sb_2O_3
250	2.0	2.29	37	0.179	2.8	3.4	0.36	16.5	9.5	1.4
260	2.2	1.73	26	0.214	2.3	3.4	0.36	13.5	5.7	1.6
260	2.6	2.13	37	0.207	2.7	4.2	0.38	16.7	8.7	1.7
290	2.4	2.28	33	0.450	2.9	3.7	0.38	22.7	11.0	0.9
265	2.3	2.09	33	0.244	2.7	3.7	0.37	17.1	8.5	1.3
6.6	12.0	14.1	18.1	51.3	10.8	10.5	3.2	23.9	33.0	29.3

Table IV. Second Matching Group of Grisaille Fragments Identified

	Percent				
Accession No.	Na_2O	K_2O	BaO	MnO	Fe_2O_3
23.229.5-4	2.5	14.1	0.135	1.01	0.51
30.73.206	2.7	14.5	0.147	1.23	0.52
30.73.207	3.1	14.9	0.127	1.08	0.51
30.73.209	2.6	15.3	0.141	1.05	0.51
30.73.213	2.6	13.6	0.146	1.03	0.50
30.73.215	4.0	12.5	0.138	1.00	0.53
Mean	2.9	14.1	0.139	1.06	0.51
Group std dev., %	19.5	7.5	5.6	7.9	2.0

tion of sodium oxide ranged from 0.5 to 3.21%, and the concentration of potassium oxide ranged from 5.27 to 8.55%. The marked difference for these two oxides between this glass and the remainder of the Rouen glass sampled (Table II) is evident.

From the random samples, two sets of compositionally related glass emerged. One set (Figure 13) is clearly stylistically related. The data for these four pieces of glass are given in Table III. The second set, composed of six compositionally matching pieces (Figure 14 and Table IV), are distinctly different in composition from the group of four. Figures 15 and 16 compare the average oxide concentration and standard deviation ranges of their variations for these two sets of glass. They show that the oxides of sodium, barium, cesium, cerium, europium, thorium, and cobalt are all significantly different in these two groups.

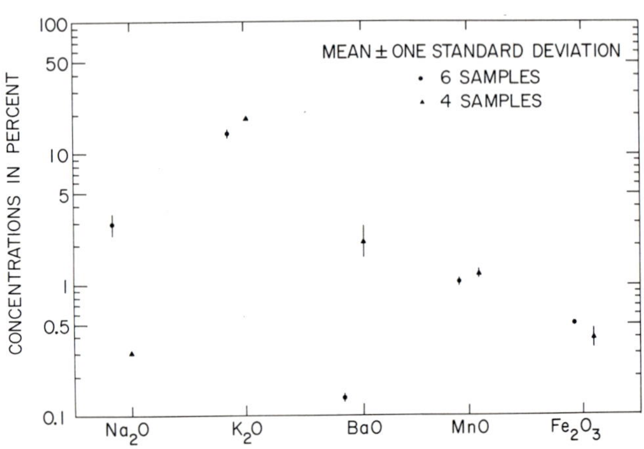

Figure 15. Comparison of mean concentrations of major and minor components in two sets of compositionally related pieces of glass from The Cloisters Museum. Specimens shown in Figures 13 and 14.

Using Accession Numbers of The Cloisters Museum (*See* Figure 2)

Parts per Million

Rb_2O	Cs_2O	Sc_2O_3	CeO_2	Eu_2O_3	HfO_2	ThO_2	Ta_2O_5	Cr_2O_3	CoO	Sb_2O_3
220	0.62	1.74	61	0.76	3.2	2.1	0.31	14.5	61	5.7
240	0.68	1.70	82	0.45	2.8	2.0	0.30	16.4	8	2.1
230	0.95	1.75	79	0.51	3.1	2.2	0.31	16.8	33	1.1
240	0.90	1.62	71	0.44	2.7	2.4	0.31	13.5	48	3.1
250	0.86	1.67	66	0.40	2.7	2.3	0.25	16.4	48	3.3
250	0.77	2.00	58	0.52	2.3	2.2	0.33	18.5	40	4.8
238	0.79	1.74	69	0.50	2.8	2.2	0.30	15.9	42	2.9
5.1	18.2	7.6	14.9	25.3	12.5	6.7	10.0	11.9	32.7	83.1

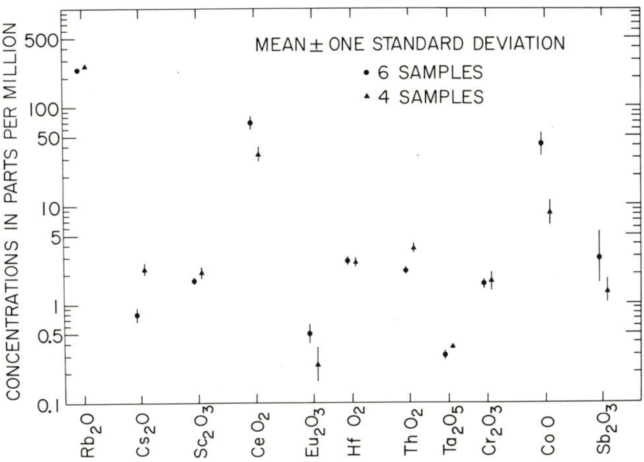

Figure 16. Comparison of mean concentrations of trace components in two sets of compositionally related pieces of glass from The Cloisters Museum. Specimens shown in Figures 13 and 14.

The matching set of six individual glass pieces is closely related compositionally to the glass of the Rouen Chateau panels. Figures 17 and 18 compare the compositions of these glasses. Obviously they cannot be distinguished on the basis of the elements determined. Thus, these six pieces of glass were probably produced by the same manufacturer as the Rouen Chateau glass.

The set of samples from a panel at the Museum of Art, Princeton University (Figure 9) have internal compositional consistencies which allow conclusions similar to those for Rouen panels. Eleven fragments from this panel were sampled and analyzed (Table V). Potassium oxide concentration was about 26%, and sodium oxide concentration was about 0.5%. This glass is completely different in composition from that

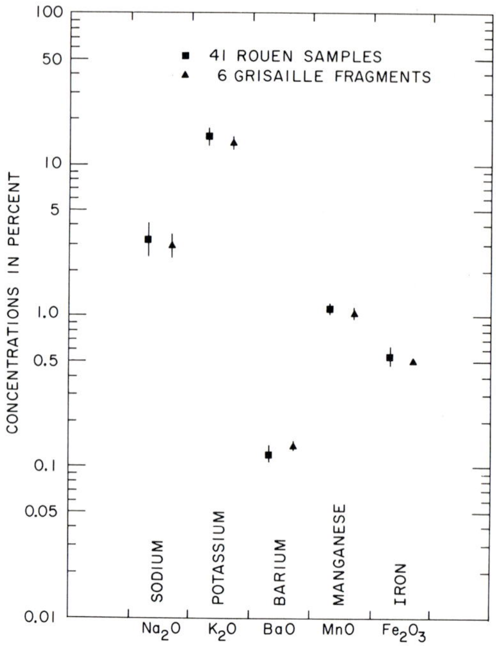

Figure 17. Comparison of mean concentrations of major and minor components in samples from the set of six matching pieces of glass shown in Figure 14 and the samples from the Rouen glass shown in Figures 2–8

Table V. Grisaille Panel—Collection of the Princeton

	Percent				
Fragment No.	Na_2O	K_2O	BaO	MnO	Fe_2O_3
1	0.57	31	0.42	1.30	0.42
3	0.51	25	0.45	1.27	0.45
4	0.55	29	0.46	1.23	0.45
5	0.56	28	0.47	1.43	0.44
6	0.53	28	0.41	1.25	0.40
7	0.54	28	0.44	1.23	0.42
9	0.49	27	0.42	1.37	0.42
10	0.54	29	0.45	1.38	0.43
Mean	0.54	28	0.44	1.31	0.43
Group std dev., %	5.1	6.4	5.0	6.0	4.1
2	0.61	21	0.36	0.94	0.58
8	0.62	22	0.33	0.86	0.53
11	0.61	21	0.36	0.93	0.52
Mean	0.61	21	0.35	0.91	0.54
Group std dev., %	0.9	2.9	2.2	5.0	6.0

of the Chateau of Rouen and all of the single pieces of glass analyzed. This lack of correlation demonstrates clearly the distinct differences in compositions observed for glasses from different sources.

This panel again shows that the composition for glass from a single source can be consistent. All specimens were so close in composition that

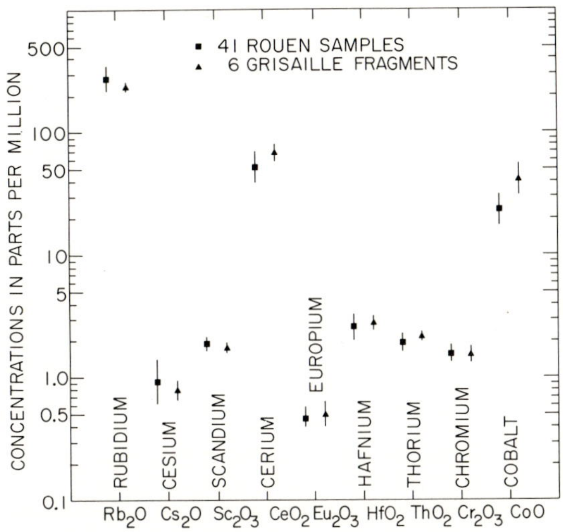

Figure 18. Comparison of trace components in samples from the set of six matching pieces of glass shown in Figure 14 and the samples from the Rouen glass shown in Figures 2–8

University Museum of Art, Accession No. 43-65

Parts per Million

Rb_2O	Cs_2O	Sc_2O_3	CeO_2	Eu_2O_3	HfO_2	ThO_2	Cr_2O_3	CoO	Sb_2O_3
590	10.0	1.38	15.2	0.28	2.2	2.7	9.6	35	1.27
580	9.8	1.38	16.4	0.28	2.1	2.0	13.8	40	1.23
630	10.2	1.50	18.3	0.30	2.1	2.2	10.3	37	1.29
640	11.2	1.48	18.7	0.22	2.2	2.3	11.4	41	1.26
610	10.6	1.33	17.9	0.20	2.3	2.0	17.6	36	0.90
600	10.0	1.41	22.4	0.25	2.4	2.2	8.7	23	1.06
620	10.2	1.37	16.7	0.23	2.2	2.0	9.1	27	1.23
600	9.8	1.39	18.5	0.25	2.4	2.1	10.3	38	1.31
608	10.2	1.40	17.9	0.25	2.2	2.2	11.1	34	1.19
3.5	4.7	4.1	12.2	14.7	5.4	10.7	26.6	24.2	13.8
834	23.2	1.60	30.3	0.28	1.9	3.4	—	22	1.97
829	26.2	1.70	27.7	0.30	3.6	2.6	12.2	19	1.20
801	27.0	1.68	17.2	0.28	2.9	2.6	13.3	21	1.51
821	25.4	1.66	24.3	0.29	2.7	2.9	12.7	20	1.53
2.2	8.4	3.3	35.6	4.1	39.8	17.2	6.3	7.2	28.2

they indicated a common origin. However, as with the Rouen glass, they subdivided into groups so similar in composition that it seems they were from individual batches of glass preparation. The data in Table V are divided into two such closely related subgroups.

All of the data in this study indicate that the compositions of what are probably glasses from different workshops tend to be distinct from one another and consistent within themselves. Also, there is often enough original glass in medieval panels to establish the characteristic composition of the glass. When data have been collected on many samples of medieval glass, it should be possible to draw conclusions about the traditions and methods of the workshops and the extent to which the products were dispersed.

Acknowledgments

We thank Jane Hayward, Associate Curator of The Cloisters Museum for permitting us to sample seven panels from the Chateau of Rouen in their collection; Frances F. Jones, Curator of Collections, the Princeton University Museum of Art for allowing us to sample a thirteenth-century grisaille panel from their collection; and The Cloisters Museum and Corning Museum of Glass for additional samples of medieval glass.

Literature Cited

1. Frankl, P., "The Chronology of the Stained Glass in Chartres Cathedral," *Art Bull.* (1963) **45**, 301.
2. Stohlman, W. F., "A Stained Glass Window of the Thirteenth Century," *Art Archeol.* (1925) **20**, 135.
3. Stohlman, W. F., "A Stained Glass Window from Chartres Cathedral," *Bull. Dept. Art Archaeol. Princeton Univ.* (Oct. 1927) pp. 3-9.
4. Stohlman, W. F., "A Window from Chartres," *The Arts* (Nov. 1927) pp. 271-274.
5. Graham, H., "A Reappraisal of the Princeton Window from Chartres," *Record Art Museum, Princeton Univ.* (1962) **21**, 30.
6. Cramp, R., "Decorated Window-Glass and Millefiori from Monkwearmouth," *Antiquaries J.* (London) (1970) **2**, 327.
7. Sayre, E. V., Smith, R. W., "Compositional Categories of Anclent Glass," *Science* (1961) **133**, 1824.
8. Geilmann, W., "Contribution to the Knowledge of Old Glasses III. The Chemical Composition of some Old Glasses, Especially German Glass, 10th-18th Century," *Glastechn. Ber.* (1955) **28**, 146.
9. Theophilus, "De Diversis Artibus," manuscript *ca.* 1123 A.D. Text and translation by C. R. Dodwell, London, 1961; translation with technical notes by J. H. Hawthorne and C. S. Smith, Chicago, 1963.
10. Dean, B., "The Exploration of a Crusader's Fortress (Montfort) in Palestine," *Bull. Metropolitan Museum Art* (New York) (1927) **22**, Part II.
11. Frankl, P., *Records Museum Historic Art* (Princeton Univ.) (1944) **3**(1), 8.

Bibliography[a]

Geilmann, W., Bruckbauer, T., "Beitrage zur Kenntnis Alter Glaser II. Der Mangangehalt Alter Glaser," *Glastechn. Ber.* (1954) **27,** 456.

Geilmann, W., "III. Die Chemische Zusammensatzung einiger alter Glaser, insbecondere deutscher Glaser des 10. bis 18. Jahrhunderte," *Glastechn. Ber.* (1955) **28,** 146.

Brill, R. H., "Scientific Studies of Stained Glass," *J. Glass Studies* (1970) **12,** 185.

Vassas, C. D., "Chemical, Thermal-Analysis, and Physical Study of Glasses of Medieval Stained-Glass Windows," *Int. Congr. Glass, Artistic Hist. Commun., 9th, 1971* (1972) 241-266 (French).

Vassas, C. D., "Colorimetric Study of Glasses of Medieval Stained-Glass Windows," *Int. Congr. Glass, Artistic Hist. Commun., 9th, 1971* (1972) 267-294 (French).

Bettembourg, J. M., "Blue Glasses of Stained-Glass Windows. Analysis by Atomic Absorption Spectrometry," *Int. Congr. Glass, Artistic Hist. Commun., 9th, 1971* (1972) 225-239 (French).

Lahanier, C., "Analysis of Stained-Glass Window Glasses by X-ray Fluorescence Spectrometry," *Int. Congr. Glass, Artistic Hist. Commun., 9th, 1971* (1972) 209-224 (French).

[a] Reported analyses of medieval stained glass.

RECEIVED July 9, 1973. Research supported in part by the U.S. Atomic Energy Commission, The National Endowment for the Humanities, and the Smithsonian Institution.

8

Neutron Activation Analysis of Gold Impurity Levels in Silver Coins and Art Objects

ADON A. GORDUS and JEANNE P. GORDUS

The University of Michigan, Ann Arbor, Mich. 48104

The levels of gold impurity in most silver coins minted more than 100-200 years ago appear to be an indication of the levels occurring in the original silver sources. Thus, it is often possible to group coins according to their gold:silver content ratios and to suggest possible geographic origins of the silver. During the past six years over 8000 coins and metallic art objects were analyzed using samples in the form of metal streaks on etched quartz tubing. Summarized here are analytical data for Sasanian (Persia: 224-650 A.D.) coins and silver art objects as well as Umayyad (Persia: 661-750 A.D.) and Byzantine coins. Data are also presented on the identification of modern forgeries of ancient and medieval coins and Sasanian art objects, the use of Mexican and South American (Potosí) silver in Spanish and other European coinage and its relationship to the price revolution of sixteenth-century Europe, and the questionable honesty in the minting of Zapatista Revolutionary Mexican coinage of 1914-1915.

Traditionally, archaeological chemistry has been restricted almost entirely to the study of problems for which the only remaining evidence was the material to be analyzed. For example, earlier studies in our laboratory which provided results that allowed identification of the geological sources of Hopewell obsidian artifacts (*1*) are what could be called classical archaeological chemistry. Must we, however, restrict archaeological chemistry to periods and areas where artifacts are the chief sources of information and avoid problems of importance in more recent times simply because written records exist? Clearly, the significance of the problem to be studied should be the criterion, and two of the three studies reported here indicate a close and productive interaction between analytical results and conventional written historical sources. In the early days of the alliance between chemistry and studies of the past it was natural for archaeological chemists to select problems for which the production of any new evidence was important. However, now, as the scientific techniques being used have become more precise, refined, and sophisticated, the attention of the investigator should turn toward problems which are subtle, much debated, and sufficiently significant to justify the intricate work involved.

Let us consider coins as our sources in the same way that historians approach their sources. Coins are of little use unless they are correctly cataloged. A great catalog, such as some of the British Museum catalogs, is, in the historian's view, equivalent to a well edited text. A poor catalog, like a poorly edited text, is often useless. Archaeological chemists can provide analyses which aid in cataloging coins, thus making these materials viable sources of information.

Very often, results of coin analysis correct false impressions easily obtained from historical and economic documents. Even allowing for wear, coins seldom meet the legal standard of weight and silver content required by the law at the time they were minted. Nevertheless, it is still surprising to find that coins from two different cities (Lucca and Pisa in the eleventh and twelfth centuries) differ considerably in weight and silver content in a period when bankers had access to assaying techniques. It is even more surprising to note the difference between their real value, as revealed by analysis, and their perceived value as revealed by documents of the period. Also, as our report on the Potosí silver will show, a great economic theorist of sixteenth-century France, whose hypotheses are influential even today in some economic circles, based his theories on what he perceived to be the case—and what we find to be, at best, a highly questionable foundation from which to proceed. It may seem that the relationship between a lithium drifted germanium gamma-ray detector and societal attitudes toward money and religion in sixteenth-century Spain is hard to establish, but the strength of archaeological chemistry has always been the productive coupling of incongruous techniques and concepts.

Experimental

Because of the nature of the samples involved in this study, special, preferably nondestructive analytical procedures must be used. In addition most museums will not permit, or at least find it difficult to arrange for, the transfer of a coin or art object to an outside laboratory. Further, the types of data to be acquired will generally be significant only if a statistically large number of samples is analyzed. These problems are all quite different from those normally encountered in analytical chemistry and thus restrict in some cases the accuracy and precision of the data.

The methods used must, ideally, permit sampling in a museum, preferably without moving the objects from the display area. The analytical methods should not be particularly time-consuming or expensive because of the volume of data generally required for valid conclusions. These fairly stringent requirements cannot be met in all cases, yet we have attempted to use them as a guide in our studies.

Analysis Methods. Until a few years ago probably no more than 2000 analyses of ancient and medieval coins had ever been performed—and then only on more common coins. The reason was simply that analysis of these artifacts required chemical methods which destroyed at least a portion of the coin. Even the less harmful method, spark spectroscopy, is not permitted by the major museums since it leaves a visible burn mark on the edge of the coin and requires temporary transfer of the coin from the museum to a laboratory.

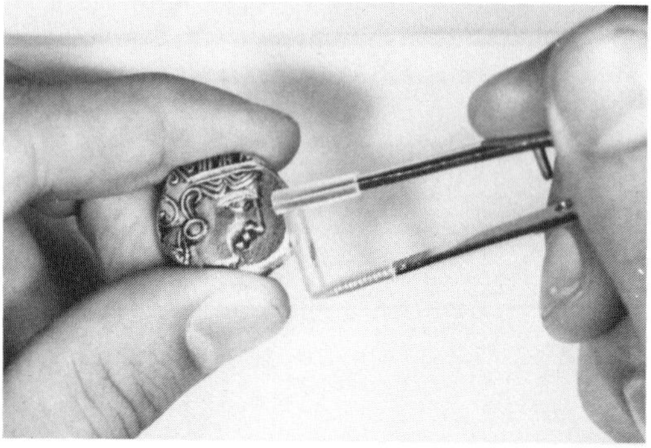

Figure 1. A small amount of non-corroded metal alloy from the edge of the coin is obtained as metal streaks on the roughened quartz tubing

Specific gravity can be used to approximate the content of gold coins but only if a binary mixture is assumed (usually Au + Ag). Density methods for silver objects are not very reliable, not only because of the necessity for assuming a binary mixture, but because the density of copper (8.94 g/ml) is relatively close to that of silver (10.5 g/ml). Possible air bubbles in the alloys (both gold and silver objects) further limit the reliability. Finally, the density method provides no information on minor or trace-level constituents.

In those few cases where a coin or art object can be brought to a laboratory, certain nondestructive analytical methods can be used. Such methods include x-ray fluorescence spectrometry and whole-object neutron activation analysis. Whole-coin neutron activation irradiation in a reactor is not feasible for large-scale studies even though a few European workers have used this method to analyze about 1000 coins. The problem arises from the production of detectable amounts of a 255-day half-life isotope of silver (^{110m}Ag). Although a single coin irradiated in this way would possess only a small amount of residual radioactivity, a museum having more than 1000 irradiated silver coins would, legally, be required to obtain an Atomic Energy Commission radioisotope-storage license.

Irradiation with charged particles accelerated in a cyclotron, while eliminating the problem of long-lived silver isotopes caused by the different nuclear reactions involved, is not feasible for any large-scale study because of the cost involved. Unlike a nuclear reactor, where numerous different samples can be irradiated simultaneously, only a single sample can be irradiated using a cyclotron beam.

Neutron Howitzer Irradiation. Whole-coin irradiation, however, has the advantage that the internal portions of the coin are irradiated and activated, thus permitting nondestructive analysis of the entire coin. One method we have devised is based on such irradiation but in a neutron beam that is about 100,000,000th as intense as that in a reactor. Only the silver in the coin is activated sufficiently for detection. None of the 255-day half-life isotope is detected, and the analysis is based on silver isotopes with half-lives of 24 sec (^{110}Ag) and 2.4 min (^{108}Ag). This silver activity dissipates in 10-15 min, and the coin is unharmed. We have analyzed over 4000 coins by this method which is described in Refs. *2, 3, 4.*

Streak Method of Analysis. Other methods of analysis, besides destructive chemical analysis, require sampling beneath the corrosion level to obtain a sample representative of the original coin. We have used a streak method of sampling (Figure 1) based on a procedure devised by E. V. Sayre of Brookhaven National Laboratory. A tiny area (about 3-4 mm^2) on the edge of a coin or metallic art object is first stroked about five or six times with fine-grain emery paper to remove about 0.03 cm of surface layer. This brightened area is rubbed with a small piece of roughened high-purity quartz tubing to produce a metal streak on the quartz. These streaks weigh less than 0.0001 mg. A second and occasionally a third streak is taken at the same position.

This method of sample taking occasionally causes consternation among metallurgists who, through experience, are familiar with the inhomogenieties which exist in metal alloys. We do not deny the existence of such inhomogenieties but simply note that the areas over which we rub the quartz tubing apparently are large enough to give consistent sets of data. These data are also consistent with similar streaks taken in the same way from other regions on the edge of the coin, indicating that there are no gross inhomogenieties in ancient and medieval coin alloys.

The streaks are brought to Ann Arbor and together with streaks from alloys of known composition are irradiated for 2 hr in the University of Michigan nuclear reactor, at 3×10^{13} neutrons/cm^2/sec, resulting in detectable amounts of radioactive silver, copper, gold, arsenic, antimony, and zinc. The activity is measured using a pulse-height multichannel gamma-ray analyzer as described below.

A more serious concern is that of corrosive depletion of the more reactive metals near the surface. Ideally, an internal drilling sample should be taken for analysis. While this is preferable to a streak sample, practical limitations are imposed by the types of objects with which we are working. We doubt if any owner or museum curator would allow us to drill into a coin. However, some museums will occasionally permit drillings to be taken from metallic art objects, and during the past year a group from Brookhaven National Laboratory (including P. Meyers and E. V. Sayre) has obtained drilling samples from silver art objects which we had previously analyzed by using streak samples. The Brookhaven group found an average of 20% more copper in the drillings. This is understandable since the reactivity of copper can result in pronounced surface depletion of that element. However, for most of our studies, this difference in copper level between streak and drilling samples is of minor concern, especially for objects with more than about 90% silver. In these cases, a streak silver content might indicate 94.0%, for example, whereas a drilling sample would show 92.9%. Since the experimental uncertainties in the calculated silver contents are typically about ±1-2%, this difference between streak and drilling data would appear minor. However, in cases where it may be important to approximate the silver content more accurately, this general difference can always be taken into account in the interpretation of the data.

On the other hand, the Brookhaven group found about 10% less gold in the drilling samples *vs.* our streak samples. As noted below, it is the gold:silver ratios that are especially important, and the Brookhaven

Table I. Analysis Data for Edge *vs.* Inner Sample Streaks[a]

Sample	*Identification*		*Ag, %*	*Cu, %*	*Au, %*	*% Cu/ % Ag*	*% Au/ % Ag (× 100)*
1[b]	Roman, Vespasian,	edge	75.8	23.9	0.34	0.315	0.44
	75 A.D.	inner	76.6	23.1	0.34	0.302	0.44
2[b]	Roman, Early Republic	edge	85.7	13.8	0.45	0.161	0.52
	205-195 B.C.	inner	83.4	16.1	0.43	0.193	0.52
3[b]	Roman, Republic,	edge	97.4	2.1	0.47	0.022	0.48
	Claudia *ca.* 41 B.C.	inner	97.4	2.1	0.46	0.022	0.47
4[b]	Roman, Titus, 80 A.D.	edge	94.6	4.6	0.73	0.049	0.77
		inner	92.8	6.5	0.68	0.070	0.73
5[b]	Roman, Severius	edge	44.0	55.4	0.20	1.26	0.45
	Alexander 222-235 A.D.	inner	43.9	55.6	0.20	1.27	0.45
6	Sasanian, Shapur I, *ca.*	edge	15.2	84.2	0.065	5.54	0.43
	250 A.D., *highly corroded*	inner	12.6	86.9	0.057	6.90	0.48
7	Sasanian, Shapur I	edge	91.8	7.7	0.50	0.084	0.55
	ca. 250 A.D.	inner	86.4	13.1	0.46	0.152	0.53
8	Sasanian, Yezdegird I	edge	96.2	3.2	0.63	0.033	0.66
	ca. 400 A.D.	inner	95.9	3.5	0.63	0.036	0.66
9	Sasanian, Khusrau I	edge	94.6	5.4	0.016	0.057	0.017
	ca. 575 A.D.	inner	96.9	3.1	0.015	0.032	0.015
10	Sasanian, Khusrau II	edge	97.6	1.8	0.57	0.018	0.58
	ca. 598 A.D.	inner	97.5	1.9	0.57	0.019	0.58
11	Sasanian, Khusrau II	edge	96.5	3.4	0.018	0.035	0.019
	620 A.D.	inner	96.3	3.6	0.019	0.037	0.020
12	Foot of a small Islamic	edge	74.9	22.7	2.4	0.303	3.2
	silver bowl	inner	75.8	22.1	2.4	0.292	3.1

[a] Negligible zinc, arsenic, and antimony were detected.

[b] Roman coins were borrowed from G. Carter who had filed down one side of the coin for x-ray fluorescence analysis. The inner samples were taken from this filed-down side of the coin. The x-ray analysis data of Carter for each of these five coins were within ±1.5% silver of that found by the streak method.

drilling samples from some of our coins listed in Table I show only a 6.3% lower value *vs.* our edge Au:Ag values. However, these same Brookhaven data are also 4.2% lower when compared with our inner Au:Ag ratios, whereas we should be in agreement for this set of data since sampling is from the same region. This discrepancy is apparently caused by differences in standards which we recently compared and found to differ in Au:Ag ratios by 4.0 $\pm$ 1.4%. This difference error must be associated primarily with our own gold–silver standards which are simple jewelers' alloys; the Brookhaven standards were prepared under more exacting requirements. Hence, all of our published Au:Ag data, including those here, should be multiplied by 0.960. The actual difference between edge Au:Ag ratios and drilling Au:Ag ratios, therefore, is probably only on the order of $6.3 - 4.2 \simeq 2\%$. (The data of Table I, for example, show that the Au:Ag ratios from the inner samples are, on the average, only 0.7% less than the edge Au:Ag ratios.)

Streak Data Calculation. Irradiated streak samples are mounted on $2'' \times 2''$ cards and placed next to the face of a high-resolution (FWHM = 2.2 keV at 1332 keV) Ge(Li) detector. Aluminum shields, 1/16″ thick, are placed on each side of the sample to provide sufficient mass to allow for reproducible ^{64}Cu positron annihilation. The gamma-ray spectral data (radioactivity counts) are determined using the following spectral peaks: ^{198}Au (2.70 day, 411.8 keV), ^{69m}Zn (13.8 hr, 438.7 keV), ^{64}Cu (12.8 hr, 511.0 keV), ^{76}As (26.3 hr, 559.2 keV), ^{122}Sb (2.75 day, 564.0 keV), ^{110m}Ag (255 day, 657.8 keV), and ^{24}Na (15.0 hr, 1368.4 keV). The ^{24}Na activity arises primarily from small impurities in the high-purity (Spectrosil) silica. However, since ^{24}Na also results in a detectable 511.0 keV peak when Ge(Li) detectors are used, the ^{64}Cu peak is corrected for the ^{24}Na, 511 keV contribution. This correction in our detection system is about 5% of the 1368 keV ^{24}Na peak and usually affects the copper data by less than 1%. In addition, ^{76}As emits another gamma-ray of lesser intensity at 657.0 keV, thus interfering with the ^{110m}Ag peak. This ^{76}As correction of the silver peak in our detection system is about 7% of the 559 keV ^{76}As peak but almost always has negligible effect on the calculated silver data because of the low arsenic contents. Only in highly debased coins, which possess higher arsenic (and lower silver) contents, does this arsenic correction of the silver data represent more than 1% of the silver activity.

Since we do not weigh these streaks, we assumed that the Ag + Cu + Zn + As + Sb + Au = 100%. By comparing the radioactivity levels in the streaks with those in the standard alloy streaks (each corrected for decay to a common reference time) we can then calculate the percent ratios: Cu/Ag, Au/Ag, Zn/Ag, Sb/Ag, and As/Ag. The combination of these ratio data and the above equation allows us to calculate the individual percent values. Various chemical analyses reported in the literature for silver coins and art objects indicate that other elements such as lead and tin (which we do not detect) are usually present at less than 2%. There are notable exceptions however. Some types of coins from certain periods contain up to 10-15% Pb + Bi. As a result, we have always performed direct neutron Howitzer silver analysis on at least a few coins of each general type that are analyzed by streak analysis. The silver data for the Howitzer analysis are invariably lower than those for the streak analysis, but this is to be expected for two reasons: the copper surface corrosion effect and the assumption in the streak analysis that Pb, Bi, Fe, and other minor elements are absent. In some cases we have simply constructed empirical correction curves (based on data for coins subjected to both Howitzer and streak analysis) to adjust the streak data for other coins of the same type. (A flameless atomic absorption spectrometric method which we plan to use will allow us to analyze the streak samples for Pb, Bi, Fe, and other, presently undetected elements.) This streak method has been described in Refs. *3* and *4*.

Gold:Silver Ratios

Examination of the gold:silver ratios from over 9000 such metal streaks shows clusterings in values that usually correlate with the mints and dates of the coins. This implies that each silver source used during each period tends to have a characteristic gold/silver level. Some geologists might object to this statement since some silver mine analysis records from the 1800's show a certain variability in the gold level of a single silver mine. However, the mine analyses were performed on the total silver + gold extracted from the ore rocks. The ancient and medieval metallurgists, on the other hand, were able to extract only certain fractions of the silver in the ore.

Since these gold impurity levels were almost always less than 1%, their presence did not color the silver visibly and therefore went undetected. Further, the chemical similarity of gold and silver resulted in the gold being carried along in the refining process, a postulate of ours that has recently been confirmed by McKerrell and Stevenson (*5*) who subjected silver ores to a lead extraction-purification process similar to that used in ancient times. The gold impurity level in a coin, therefore, can be a direct reflection of the level in the silver in the original ore, and coins can be grouped according to these gold levels. The interpretation of such gold-level data is a little more complicated since coins and art objects were made not only of new silver but sometimes of varying amounts of silver obtained from melting older coins or art objects.

It has been implied (*6*) that the only silver available within the Persian Empire was that present as an impurity in lead ores. The extraction of silver from lead ores was certainly practiced by the Romans, especially when high-grade silver sources became depleted. However, the fact that such extraction technology was known does not mean that the Persians, for example, necessarily used this method, especially if high-grade silver ores or metallic silver sources were still available. In fact, Arab chronicles, even those of the tenth through thirteenth centuries, clearly distinguished lead sources, silver sources, and sources from which both lead and silver were derived (*7, 8*). The phrase "silver source" as used in this article is meant to imply a geological formation that was viewed by the people of the period primarily as a source of silver, as compared with a formation used, for example, primarily as a source of lead. This definition, therefore, includes metallic silver sources, argentite (Ag_2S), cerargyrite ($AgCl$), and other mineral formations containing

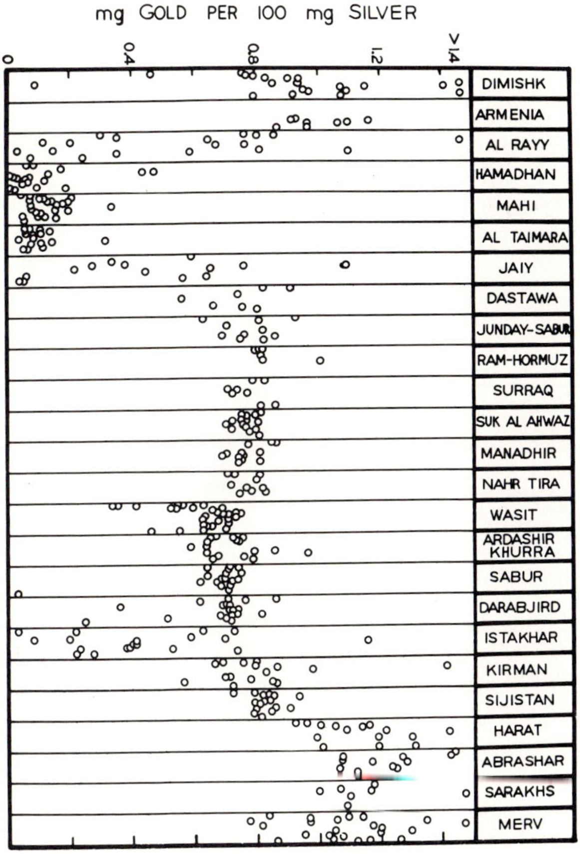

Figure 2. Gold impurity levels in Umayyad silver dirhems for 25 mints issuing coins during the period A.H. 90-98/708-717 A.D.

significant amounts of silver. Included would be those galena ores having high silver (argentite) contents, but typical galena ores having silver:lead ratios less than perhaps 1:10 (most contain much less silver) would be classed as lead ores.

Umayyad Coinage

During the nine-year period 708–717 A.D. (which corresponds to the Islamic dates 90–98 A.H.), coins were issued by many Umayyad mints. Figure 2 shows gold impurity level data for 378 Umayyad coins issued at 25 mints during this period. These data are arranged geographically (Figure 3) beginning at Damascus, going northeast to Armenia, southeast to Al-Rayy, and then south and curving back north to Merv. Several data clusters are apparent in Figure 2. The mints of Hamadhan, Mahi, and Al-Taimara apparently relied almost entirely on a silver source (or sources) having very low gold levels. Junday-Sabur, Surraq, Suk Al-Ahwaz, Manadhir, and Nahr Tira appear to have used a silver source having about 0.75% gold impurity whereas Wasit, Ardashir-Khurra, Sabur, and Darabjird relied mainly on a source with a gold-in-silver level of about 0.70%. The Istakhar data exhibit a range of values suggesting the use of mixtures of silver including at least one which had low gold. This low-gold source cannot be the same one used by Hamadhan, Mahi, and Al-Taimara since the use of silver with low gold is not apparent in the coinage of geographically intermediate cities such as Sabur and Ardashir-Khurra. Sijistan and to some extent Kirman used

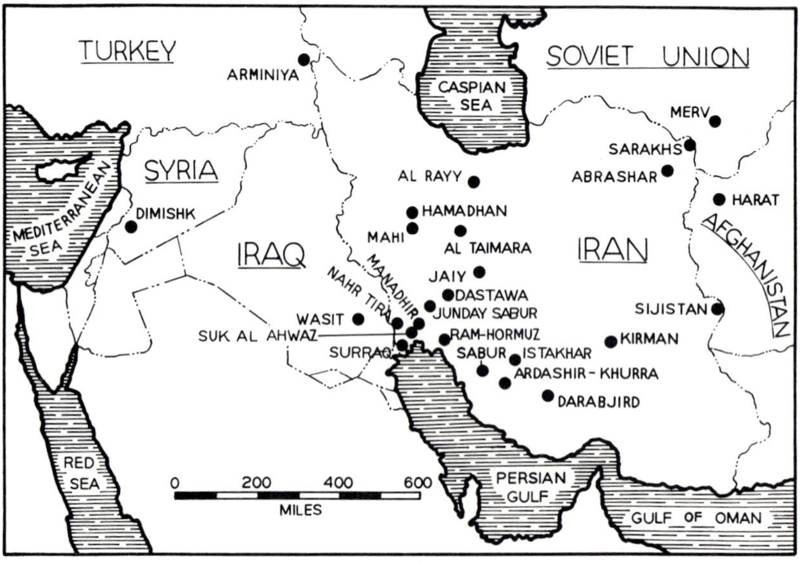

Figure 3. Locations of Umayyad mint-cities for which data are given in Figure 2

silver having about 0.80% gold, and the northern mints of Harat, Abrashar, Sarakhs, and Merv apparently utilized silver from various sources, most of which tended to have more than 1.00% gold impurity.

Equally interesting are the important cities of Rayy and Jayy, both of which were on principal trade routes. It is not unreasonable, therefore, to find that the range of data includes gold levels associated with cities to each side of Rayy and Jayy. Similarly, coinage from the city of Wasit, as well as a few others such as Darabjird and Istakhr, show use of multiple silver sources. It thus appears that at least six silver sources (two with low gold, one each with 0.70, 0.75, 0.80, and >1.00% gold) were exploited by the Umayyads. One must have been very close to the Hamadhan, Mahi, Al-Taimara region. Similarly, the location of some of the remaining sources can be approximated. The 0.75% gold-in-silver source must have been very near the four cities of Surraq, Suk Al-Ahwaz, Manadhir, and Nahr Tira. The Southern mints of Darabjird, Sabur, and Ardashir-Khurra must have been near the 0.70% gold impurity source whereas Sijistan and Kirman shared a source having silver with about 0.80% gold. The data for each of the four mints in the Northeast exhibit too large a range of values to permit suggesting the general location of their silver sources, and the spread of these data suggests the usage of more than one silver source. Because of the large distance between Damascus and Armenia and the other Umayyad mint cities, it is difficult to draw conclusions from the data for the coinage of these two Western mints.

Data such as these can help answer certain historical questions. For instance, the mint city written on Umayyad coins called Mahi must be the one noted here and not a similarly named city near Merv, which had been suggested as an alternative choice (9), since the gold impurity data obtained from coins minted at Mahi agree with the data for Hamadham and Al-Taimara but do not agree with the data for Merv. By determining the types and amounts of silver used in each city, it may be possible to assess the movement of silver throughout the Empire. Furthermore, possible locations can be suggested for ancient silver sources. This could have important geological implications. For example, the Geological Survey of Iran does not list a single silver source in all of Iran, and it is usually stated that there never were any silver sources in Iran. (The only silver found today in Iran originates as a trace-level impurity in lead ores; we have determined that the gold-silver ratio data for these ores is much less than that found for any of these coins.) The data of Figures 2 and 3, however, suggest that during the Umayyad period there probably was a silver source in the Hamadhan, Mahi, Al-Taimara region. This region, therefore, could be appropriate for present-day geological exploration. Possibly even the silver mine used during the Umayyad period could be worked today using modern extraction methods.

Sasanian Coinage

Silver Fineness. Figure 4 contains data for the silver fineness of about 400 Sasanian coins as determined by the direct (neutron Howit-

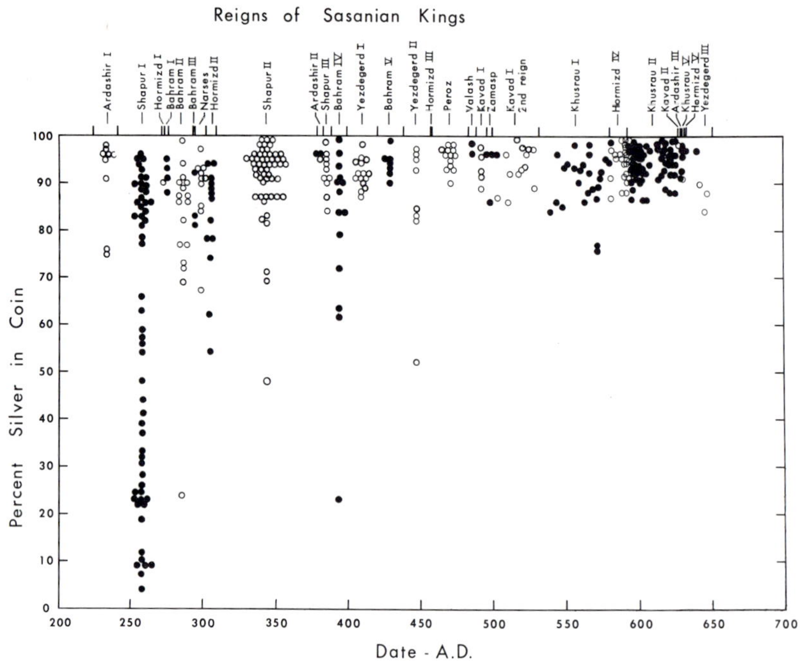

Figure 4. Silver fineness as determined by neutron Howitzer analysis for about 400 Sasanian drachms. Alternating open and closed circles are used to assist visually in associating data points with ruling monarchs listed at top of graph.

zer) coin irradiation method. Alternating open and closed circles were used to help visually identify data points according to the ruling monarch. This graph is an updated, expanded version of the one recently published together with a preliminary discussion of the data (*10*). Recent emphasis has been placed on the analysis of the coins of low-silver fineness, particularly during the reign of Shapur I. Since these debased coins are seldom found in extremely fine condition, they are underrepresented in museum collections. As a result, a number of Sasanian coins that looked debased were bought from dealers in France, England, and the Near East and subjected to both Howitzer and streak analysis. These various data, and especially those for the early coinage (Ardashir I–Shapur II: 224–379 A.D.) are discussed in detail in a separate article (*11*).

Gold Impurity Levels. Shown in Figure 5 are gold impurity data for approximately 1300 Sasanian coins with almost half of these data points representing analyses of Khusrau II coinage. Unlike Figure 2, these gold level data are plotted on a logarithmic scale to show subtle differences in the low-gold data points. Only during the later Sasanian period do the coins bear mint monograms and dates given in terms of the regnal year. Identifications, where necessary, were based on Göbl (*12*). These data suggest that various silver sources were exploited, and most of these sources had more than about 0.4–0.5% gold impurity except during the later reigns when silver of low gold levels was occasionally used.

Preliminary examinations of the gold levels in Khusrau II coinage grouped according to mint monograms indicated that the Pahlevi mint signatures usually identified as ZD and RD, frequently considered variants both representing Rayy, were actually associated with two separate mint cities (*4*). It was also found that MR and MB monograms represented two separate mint cities and were not simply variants of a Merv signature (*4*). Work of this type is continuing and is being extended to the reigns of Hormizd IV and Khusrau I.

The data represented by the open squares in Figure 5 were obtained from streaks of modern fake Sasanian coins (*4*). Note that the gold levels for these fake coins are usually appreciably less than those for authentic coins. This is because silver used in the last 100 years almost always

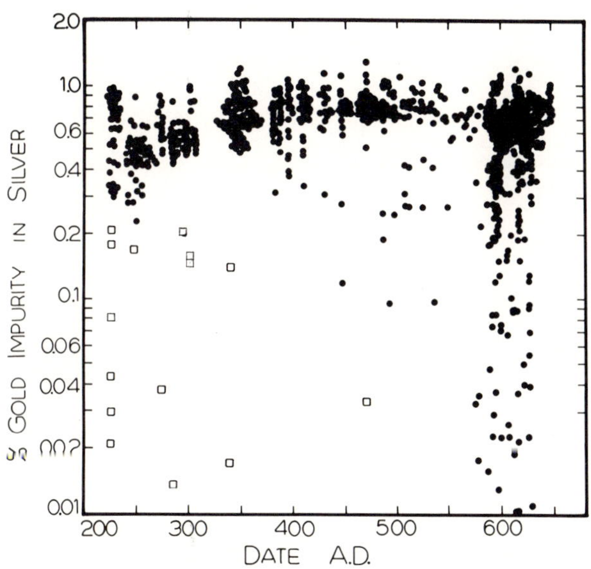

Figure 5. Gold impurity levels in over 1300 Sasanian drachms. Open squares are gold levels in modern fake Sassanian coins.

had less than about 0.1% gold; present-day silver usually has less than 0.01% gold since modern refining methods allow economical removal of most trace-level gold.

Sasanian Metallic Art

A number of silver plates, bowls, and pitchers have been identified as being made in the Persian Empire during the Sasanian period (*13, 14,*

15). Recently, some art historians have reassigned the manufacture of a few of these objects to slightly later periods or adjacent geographical regions. Some of the objects, particularly those in the shape of plates, show a king (frequently on horseback) hunting wild game. Almost all art historians still attribute these particular motifs to the Sasanian period and region.

These *objets d'art* are rare and command a high price. The increased demand by museums and private collectors for metallic art of this period has been accompanied by an increase in the supply of these objects—and by an increasing suspicion on the part of art historians, museum curators, private collectors, and reputable dealers that some (and perhaps many) of these silver Sasanian art objects are modern forgeries. Virtually all art historians agree that the Sasanian objects in the State Hermitage Museum (Leningrad) are authentic, with one exception—a late nineteenth century copy of a plate in the British Museum. This agreement is based to some extent on the fact that the Leningrad collection has a suitable "pedigree." It includes objects that either were found in controlled archaeological excavations or were in Russian collections as early as the mid-1800's.

The present-day art market is such that a pedigree is rare. If a purported Sasanian silver art object is put up for sale, the circumstances surrounding its origin are usually vague. If it is a fake, the vagueness is understandable, but a vague origin would also be ascribed to an authentic piece since it is illegal to export such objects from Iran, where they are generally claimed to be found near the Caspian Sea by "peasants digging in the hills."

Through the kind and willing cooperation of museum directors, private collectors, and dealers both in the U.S. and abroad it was possible to take streaks of hundreds of such silver Sasanian and Sasanian-styled art objects. Included were the entire Leningrad collection and also numerous privately owned "Sasanian" plates that virtually every art historian of this period has described as either decidedly suspicious or probably fake. The streak analysis data for those plates involving a Sasanian king–hunting scene are given in Figure 6.

These data are identified according to the general consensus of art historians and curators regarding the authenticity of each plate. As seen in Figure 6, all plates for which there is unanimous or near-unanimous agreement about their authenticity have % Au/% Ag ($\times 100$) values in excess of 0.60, except two. The first is a plate in the Hermitage museum (Au in Ag $=$ 0.18) showing Shapur II on horseback (*16*); the style in which the figure, horse, and lions are depicted differs from that on other plates in the Leningrad collection. It is possible that this object has been either misattributed temporally or geographically. The second is a plate

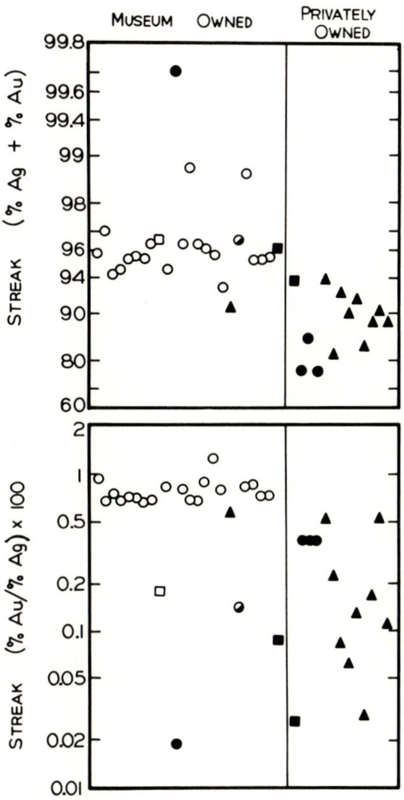

Figure 6. Streak-analysis data for Sasanian-style silver plates with a design depicting a king (frequently on horseback) hunting wild game. Art historian and museum curator general consensus: ○ = authentic; □ = authentic, but possibly misattributed; ◕ = some question regarding authenticity; ● = fake copy of a museum plate; ■ = fake solid plate; ▲ = fake double-walled plate.

(*17*) in the Tehran Archaeological Museum for which a set of streaks taken from the edge contained Au in Ag = 0.14.

Figure 6 also shows that all except one of the plates that are agreed to be authentic have Ag + Au streak values greater than 94%. The exception is a cast plate at the Bibliothèque Nationale (*18*) with 93.2% Ag + Au. In addition, none of the agreed authentic plates are double-walled. Although many have high-relief figures, these features are generally created by the addition of (cast) portions of the figures.

On the other hand, many of those plates generally considered modern fakes were made by a repousée technique and involved an outer, plain-curved back that was folded over at the rim to form a double-walled hollow plate. Occasionally a separate (cast) piece, usually a raised front leg of the horse, was attached, probably by soldering. These fakes and copies of existing museum plates all have less than 0.6% Au in Ag (in agreement with the fake Sasanian coins, Figure 5) and 14 of the 18 fakes in Figure 6 have less than 94% Ag + Au.

Similar comparisons regarding streak % Ag + Au, streak % Au in Ag, and the lack of a double-walled construction technique apply to all other objects that are agreed to be authentic and of Sasanian origin. These include plates with other designs such as dancing girls, certain birds, or animals.

These observations taken together can suggest criteria to use in judging the possible authenticity of a Sasanian silver art object: (a) The streak Au in Ag value should be greater than about 0.4–0.6%. (b) The streak Ag + Au value should be greater than about 94%. (c) The plate or bowl should not be double-walled. If none of these criteria is met, there is a strong possibility that the object is a fake. It should be stressed that these three criteria are based on the comparison between sets of data such as those shown in Figure 6. The criteria are not absolute in the way that ^{14}C or thermoluminescent dating can provide direct evidence of age, and some exceptions may exist. However, since today's cost of such *objets d'art* can frequently exceed $100,000, museum curators must exercise caution in making purchases, especially since (in the United States) the funds for such purchases invariably are derived from income tax-exempt donations. The same caution must be exercised by a museum in accepting a gift of such an object since again a tax deduction is involved.

Byzantine Miliaresia

In a joint project with D. Michael Metcalf of the Ashmolean Museum, Oxford, a series of streaks from eighth-century Byzantine miliaresia were analyzed to determine the levels of gold impurity in silver. A few of these coins were overstruck on Islamic dirhems as seen by the presence of faint remaining traces of Arabic inscription. These miliaresia generally had much lower gold impurity levels, in accord with the late Umayyad and early Abbasid coinage. From these analyses it has been possible to suggest which of the other miliaresia may also have involved overstrikes and thus estimate the fraction of silver in Byzantine coinage that was obtained from Arabic sources. This study will be described in detail in a separate article (*19*).

Hispanic-American Silver and the "Price Revolution" of the Sixteenth Century

Soon after Cerro Rico, the silver mountain discovered by the Spaniards in 1545 at what is now Potosí, Bolivia (*20*), began to produce massive amounts of the white metal and as the Spanish ships arrived to deposit the treasure in the Casa de la Contratacíon in Seville, Azpilceuta and other Dominicans of the School of Salamanca warned of the potential inflationary effect of this silver on the Spanish economy (*21*). Jean Bodin, who might well be called the first important exponent of the quantity theory of money, held, further, that all Europe was affected adversely by the treasure of the Indies, which, he maintained, was the cause of "la vie chère" (*22*). More recently, Earl J. Hamilton, in a monumental study correlating the arrival of the silver and the rise of prices in Andalusía and other Spanish provinces, concluded by assigning to the Spanish-American silver a massive role in the price revolution of the sixteenth century (*23*).

Although the Hamilton thesis has been effectively challenged on a number of fronts (*24*), the precise role of this American silver in the European economy has not been firmly established. In the absence of evidence to the contrary, it has often been assumed that Potosí silver infiltrated virtually every currency in Europe and points East with dramatic effects (*25*). Emphasis has been placed on the inflationary phenomenon itself, considered to be a result of the silver import (*23*). Since (a) it is very difficult to assess the amount of non-Potosí, (*i.e.*, European

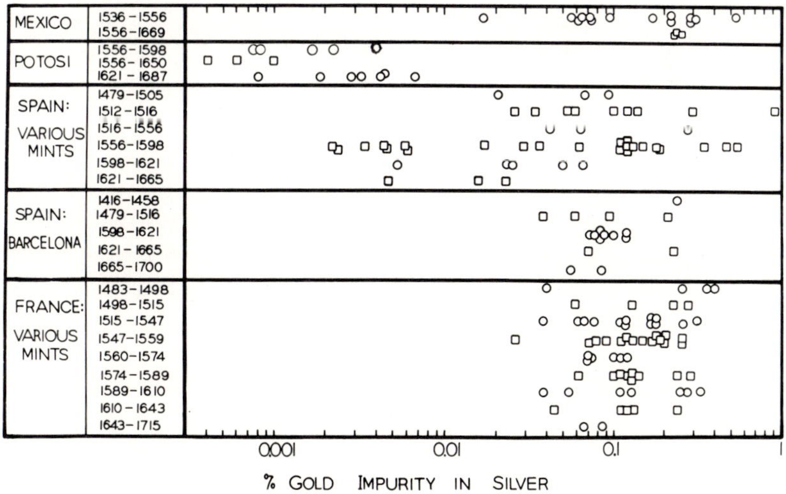

Figure 7. Gold impurity levels in Mexican, South American (Potosí), and European coinage

and Mexican) silver available for coining during the period, (b) additional complications arise because the abundant archival evidence for the silver mine production at Potosí has not yet been explored, and (c) there is evidence that "leakage" occurred (*26*) diverting some American silver from its Seville destination, it has usually been assumed that Potosí silver infiltrated the currencies of Europe. It is this basic assumption to which the preliminary study was addressed.

Fortunately, for our purposes, the silver from Potosí has a characteristically low level of gold impurity which allows it to be differentiated clearly from the Mexican-European silver stock. As shown in Figure 7 (*27*), silver from this latter stock was used in Spain, Barcelona, and France, before 1570–1580, when the Potosí mine began to be exploited on a large scale and its yields shipped to Spain.

Potosí silver was certainly used in Spanish coinage; in fact, the crudely struck réales bearing the Potosí mint mark were legal tender in Spain although much of the silver was reminted into Spanish réales of a higher artistic standard. However, none of the Potosí silver appears directly in the coins of Barcelona or France. We might have expected, as some scholars have (*28*), that at least some French coinage could be made of Potosí silver on the grounds that so much more money was minted in France in that period than before; however, it is now clear that evidence for a direct causal link between Potosí silver and increased French mint output cannot be adduced. Particular attention was paid to coins from such port cities as Marseilles and Bayonne, to check both French coastlines where silver might have "leaked" in the sense that these might have been convenient places for illegal transactions. In the same way, those mint towns on the route overland through France, where American silver was shipped by land to the Netherlands from Spain, were also investigated. Over 150 carefully selected French coins (including those for which data are given in Figure 7) were analyzed, and no Potosí silver was identified.

Despite the long-standing antagonism between Castile and Catalonia, it might have been expected that since Barcelona was the point of departure for the silver from Spain to Genoa where it was exchanged for gold and where business was transacted with Genoese bankers, Barcelona might have received some Potosí silver and minted it. In fact, it seemed so likely that only 25 coins of Barcelona were analyzed, concentrating on the likeliest dates, but again no Potosí silver was found. However, the Genoese did mint some coins made of Potosí silver. Milan, another banking center, and perhaps more importantly a city noted for arms manufacture, received Potosí silver, as shown by 5 of the 40 coins analyzed. This is still a very small amount. The enterprising English seamen who captured Spanish galleons brought the silver to England, and vast quantities of it are supposed to have arrived at the Tower mint. Again, however, analysis of 60 Elizabethan coins chosen to coincide with such deliveries reveal no Potosí silver.

Hamilton assumed that a large proportion of the Spanish treasure went to Germany, probably mostly to pay the important bankers to the Spanish Crown, the Fuggers of Augsburg. Coins of Germany are yet to be analyzed, but, except for this German destination, the results so far obtained lead us to question whether the American treasure had any real connection at all with the very real European inflation. We do not assert that the American treasure had no effect on the European economy; we question whether it was real silver which caused money to become cheap or whether it was the expectation of silver.

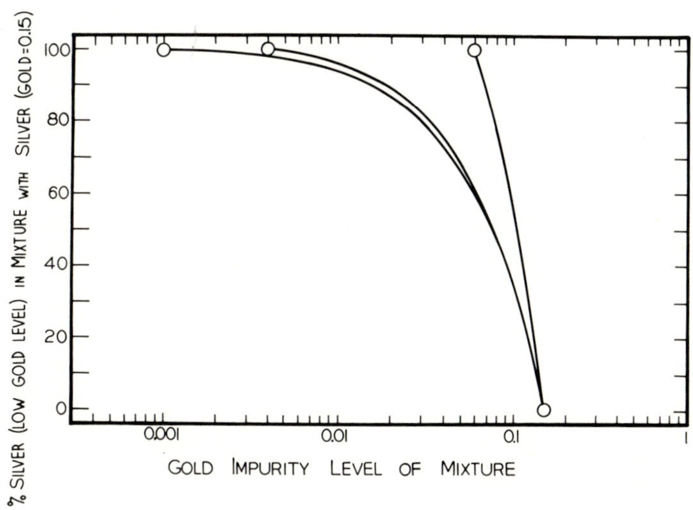

Figure 8. Percent silver having a low gold level (0.001, 0.004, or 0.06% Au in Ag for each of the three curves) in a mixture with silver having 0.15% Au in Ag impurity as a function of the resultant gold impurity level of the mixture

This study has thus far been based on the analysis of a single diagnostic element: the gold impurity level. As a result, we are not able to distinguish Mexican and European silver. Equally important, we are not able to determine to a high degree of certainty which, if any, of the European coins we analyzed might be made of mixtures of Potosí and other silver. It should be emphasized that because of the low gold impurity level in Potosí silver, the addition of small amounts of silver having large gold impurity levels could markedly change the overall gold level of the mixture. As shown in Figure 8, a mixture made, for example, of 20% European silver (gold level = 0.150%) and 80% Potosí silver (gold level

= 0.004%) has an overall gold impurity level of 0.033%. Therefore, some of the European post-1550 coins could possibly involve mixtures of which Potosí silver is a predominant component. One such coin, a French Teston of Henri II (1547–1559) having a gold level of 0.024 (data point shown in Figure 7), could be of this type.

Clearly, the use of gold as the only diagnostic element is inadequate to distinguish pure Potosí silver and some mixtures containing Potosí silver and does not allow us to distinguish pure Mexican and pure European silver. Data for other elements, which can serve in the same manner as gold, are needed to distinguish these alternative possibilities. To accomplish this, methods are presently under investigation whereby the streak sample would be dissolved following neutron activation analysis and the dissolved sample would be subjected to flameless atomic absorption spectrometry. The initial results indicate that the lead and bismuth contents of the dissolved streak samples can be detected easily, and these additional elemental data may serve partially for diagnostic differentiation. It should also be possible to detect Cd, Fe, Zn, Al, and perhaps other metals by this procedure.

The Silver Coinage of Zapata, 1914-1915

The Mexican Revolution caused enormous economic dislocation, one aspect of which was the collapse of the monetary system. The Mexico City mint, which had suspended gold coinage in 1910, ceased to strike the pesos fuertes in March 1914 and the subsidiary silver in September. Federal gold and silver coins soon disappeared from circulation. Throughout the country paper money of irregular circulation and precarious value was issued. It was widely counterfeited, and even the genuine paper might be declared invalid overnight. Some coins were produced locally and unofficially—as far as the federal government was concerned—in an attempt to stabilize values and provide small change. These coinages exist in great quantity and disparity, differing from the Federal and from each other in alloy, weight, types, legends, and denomination.

Zapatista silver one- and two-peso pieces, Figure 9, were first struck in Guerrero in 1914; fractional bronze did not appear until the following year. The two silver denominations were proportionately smaller and contained less weight of silver than the previous federal coinage, but their acceptance was encouraged by the announced presence of a stated amount of gold in each. The peso bears the legend "ORO: 0,300 {grams}," the two pesos, "ORO: 0,595."

A study of this coinage, which was initiated by T. V. Buttrey, has been described (*29, 30*). By careful examination of a number of the existing coins Buttrey was able to determine that almost all of the dies

Figure 9. Zapatista one- and two-peso coins issued in 1914 and 1915 and patterned after the previous federal coinage

used to mint the coins were cut by the same hand. By arranging the coins in terms of imperfections in the lettering caused by wear and partial breaking of the letter fonts used to prepare the dies, he was able to arrange sequentially the order in which the 1914 and 1915 coins were minted.

Combined neutron Howitzer and streak analyses of 125 one- and two-peso coins not only reveal the real gold content of these issues but help to establish the silver standard, which was probably aimed at 900 thousands fine (90% silver), a creditable alloy and better than the 800 thousands (80%) of the federal subsidiary coinage. It was therefore possible to estimate fairly closely the theoretical composition and value of the Zapatista silver coinage.

The coins reveal certain aspects of Zapatista monetary policy and control. First, an unprecedented alloy was created from which, it was calculated, coins equal in intrinsic value to the federal could be produced. However, this theoretical system was not consistently practiced. The major 1914 issues of the two pesos generally kept to a high gold content, but two minor mints and all mints in 1915 did not. As shown in Figure 10, the average silver and gold contents, while equal to or even exceeding the peso value in the early coin, showed marked debasement or outright malfeasance in the minting of some of the later issues, and especially in the one-peso issues.

The proclaimed gold content represents only about 2% gold since the value-ratio of 1 gram of gold *vs.* 1 gram of silver was about 27:1. As noted above, many silver sources contain gold as an impurity. For example, the data for sixteenth-seventeenth century Mexican silver pieces-of-eight and lower denomination réales shown in Figure 7 indicate about 0.07–0.3% Au in Ag. Therefore, the Zapata coins having higher gold contents that the proclaimed values were probably made by using new silver which already contained some gold impurity; its presence, however, was not realized by the minters, and the actual proclaimed amount of gold was added, in effect, to the gold already present in the silver. On the other hand, those one-peso pieces having gold contents much lower than ORO: 0,300 must also have been made of new silver. However, in

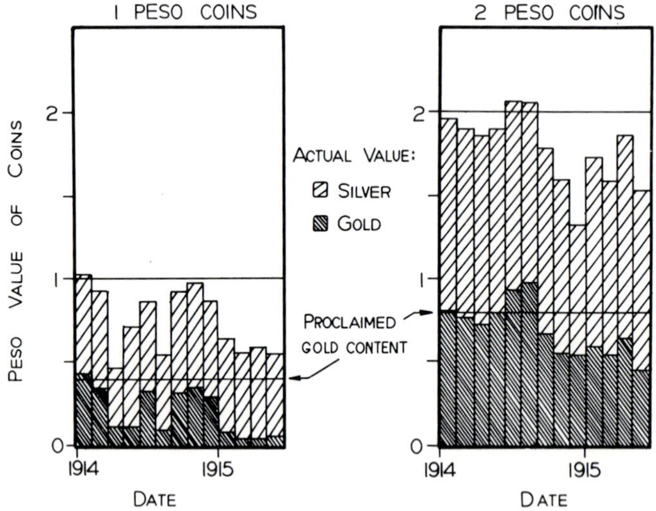

Figure 10. Proclaimed gold contents (in peso equivalents) and actual average silver and gold content values for Zapata one- and two-peso coins

this case the intent was to deceive and, in fact, no gold was actually added to the silver. The detected gold level is simply the impurity in the silver used for the coinage.

It would be easy to draw a cynical conclusion from these data which show that coins of lower denomination—namely those most likely to be used by the poorer people—were debased to a greater degree than the two-peso pieces which were used in major transactions by the revolutionary groups and were known in the American border states as the Guerrero dollar. But care should be taken to dissociate these apparently self-serving economic manipulations as shown through the coinage from the main thrust of the Zapatista revolutionaries. Zapata's coalition was composed of two socio-economic groups (*31*). The first was composed of disaffected intellectuals who supposedly had urban ties; it is most unlikely that these intellectuals were minters. The other forces were peasants whose group resources allowed them to promote their own special political interests. However, even a relatively affluent peasant is unlikely to have engaged in the financial part of the revolution. In fact, the limitation of the Zapatista movement was that it was a peasant movement; troops marched out from villages for combat and returned to farms after the fray. If we are to judge what appears to be a betrayal of the interest of the people by the higher eschelons in Zapata's revolutionary cadre, we must recall that we are dealing with a monetarily unsophisticated group that may very well have enlisted the services of moneyers not originally Zapatistas and that remained unaware of the debasement of the coins.

When we consider that the Zapatistas had homemade arms and no organized system of supplies, it seems unlikely that many of them realized their money was debased. Clearly, they never had much of it for when they made their victorious entry into Mexico City, the Zapatistas had no food and had to beg from householders.

The study of revolutions has always produced instances of counter-revolutionary and opportunistic behavior which are then used as evidence for the bad faith of all the participants or at least their leaders. The worst charge in the light of the evidence presented here (as well as the other sources cited) which can be leveled at Zapata, or for that matter at his Northern counterpart, Pancho Villa, is that neither had sufficient political or economic skill to govern the country. Both adamantly refused the office of president and felt themselves ill-equipped to provide anything more than a temporary caretaker arrangement. So the famous sign over the Palacio Nacional advertising the lack of a potential president of a high caliber: "Wanted—an honest man" (*32*) should not be taken together with this monetary evidence to demonstrate moral bankruptcy of the Mexican Revolution. If a lesson must be drawn, it should be that men of such limited means could change the face of a nation and yet remain humble enough to realize their own shortcomings, and even make this self-appraisal known.

Acknowledgment

None of the studies reported here would have been possible without the gracious and willing cooperation of curators both in the United States and Europe. George C. Miles of the American Numismatic Society, T. Chase of the Freer Gallery, P. Harper of the Metropolitan Museum, D. M. Metcalf of the Ashmolean Museum, N. Lowick of the British Museum, R. Curiel of the Bibliothèque Nationale-Cabinet des Medailles, and V. Lukonin, B. Marshak, and E. Zeymal of the State Hermitage

Museum, Leningrad, are among those who allowed us to take streaks from coins and art objects in their collections. The University of Michigan, through the Department of Chemistry and the Michigan Memorial-Phoenix Project, as well as the U. S. Atomic Energy Commission, through its Division of Research (of which this is report COO-912-27), provided the research funds which made this study possible. Through the years numerous undergraduate science students assisted in the analyses, and without their aid these thousands of analyses would not have been possible.

Literature Cited

1. Gordus, A. A., Griffin, J. B., Wright, G. A., "Activation Analysis Investigation of the Geologic Origins of Prehistoric Obsidian Artifacts," Chap. 16, pp. 222-234, "Science and Archaeology," R. Brill, Ed., MIT Press, Cambridge, 1971.
2. Gordus, A. A., "Quantitative Non-Destructive Neutron Activation Analysis of Silver in Coins," *Archaeometry* (1967) **10**, 78-86.
3. Gordus, A. A., "Rapid Non-Destructive Activation Analysis of Silver in Coins," Chap. 10, pp. 145-155, "Science and Archaeology," R. Brill, Ed., MIT Press, Cambridge, 1971.
4. Gordus, A. A., "Neutron Activation Analysis of Coins and Coin-Streaks," "Methods of Chemical and Metallurgical Investigation of Ancient Coinage," Royal Numismatic Society, *Spec. Publ.* **8** (1972) 127-148 + Plate XV.
5. McKerrell, H., Stevenson, R. B. K., "Some Analyses of Chemical and Metallurgical Investigation of Ancient Coinage," Royal Numismatic Society, *Spec. Publ.* **8** (1972) 195-210.
6. Wertime, T. A., "A Metallurgical Expedition through the Persian Desert," *Science* (1968) **159**, 927-935.
7. LeStrange, G., "The Lands of the Eastern Caliphate," University Press, Cambridge, 1905.
8. al-Kazvini, H. A. M., "The Geographical Part of the Nuzhat-al-Qulub," tr. by G. LeStrange, *E. J. W. Gibb Mem. Ser.* (1919) **XXIII.2.**, 193.
9. Walker, J., "A Catalog of Arab-Byzantine and Post-Reform Umaiyad Coins," p. lxxxviii, British Museum, London, 1956.
10. Bacharach, J. L., Gordus, A. A., "The Purity of Sasanian Silver Coins," *J. Amer. Oriental Soc.* (1972) **92.2**, 280-283.
11. Gordus, A. A., "Non-Destructive Analysis of Parthian, Sasanian, and Umayyad Silver Coins," in "Near Eastern Numismatics, Iconography, Epigraphy, and History: Studies in Honor of George C. Miles," pp. 141-162, D. K. Kouymjian, Ed., American University of Beirut, Lebanon, 1974.
12. Göbl, R., "Sasanian Numismatics," Klinkhardt and Niermann, Braunschweig, (English ed.), 1971.
13. Lukonin, V. G., "Archaeologia Mundi: Persia II," Nagel Publishers, Geneva, 1967.
14. Ghirshman, R., "Persian Art: 249 B.C.–A.D. 651," Golden Press, New York, 1962.
15. Orbeli, J., Trever, C., "Orfeverie Sassanide," Moscow and Leningrad, 1935.
16. Lukonin, V. G., *op. cit.*, Figures 136, 138; Ghirsham, R., *op. cit.*, Figure 253; Orbeli, J., Trever, C., *op. cit.*, Figure 6.
17. Ghirsham, R., *op. cit.*, Figures 248, 249, 250.
18. *Ibid.*, Figure 252.

19. Gordus, A. A., Metcalf, D. M., "The Alloy of the Byzantine Miliaresion and the Question of the Reminting of Islamic Silver," in press.
20. Porteous, J., "Coins in History," pp. 170-171, Plate 194, Weidenfeld and Nicolson, London, 1969.
21. For valuable elaborations on the Spanish economic theoreticians, *see:* Grice-Hutchinson, M., "The School of Salamanca," Oxford University Press, 1952, and Vilar, P., "Les primitifs espagnols de la pensée economiques," in "Mélanges offerts à Marcel Bataillon," Colin, Bordeaux, 1963.
22. Bodin, J., "Response à M. Malestroit," Henri Hauser, Ed., Colin, Paris, 1932.
23. Hamilton, E. J., "American Treasure and the Price Revolution in Spain," Harvard University Press, Cambridge, Mass., 1934.
24. Hammarström, I., "The Price Revolution of the Sixteenth Century, Some Swedish Evidence," in "The Price Revolution in Sixteenth Century England," collected by P. H. Ramsey, Methuen, London, 1971.
25. Wiebe, G., "Zür Geshichte der Preisrevolution des XVI und XVII Jahrhunderts," Vol. II, Leipzig, 1895.
26. Braudel, F., "The Mediterranean and the Mediterranean World in the Reign of Philip II," Vol. I, p. 476, transl. by S. Reynolds, Collins, London, 1972.
27. Gordus, A. A., Gordus, J., LaDurie, E. L., Richet, D., "Le Potosi et la Physique Nucleaire," *Ann.: Econ. Soc., Civilis.* (1972) 1235-1256.
28. Spooner, F. C., "L'Economie Mondiale et les Frappes Monetaires en France, 1493-1689.," C.N.R.S., Paris, 1960.
29. Buttrey, T. V., Gordus, A. A., "The Silver Coinage of Zapata: 1914-1915," *Hispanic Amer. Hist. Rev.* (1972) **52,** 456-462.
30. Buttrey, T. V., Gordus, A. A., "A Neutron Activation Analysis of the Silver Coinage of Zapata, 1914-1915," *Amer. Numismatic Soc. Museum Notes,* in press.
31. Wolf, E. R., "Peasant Wars of the Twentieth Century," pp. 29-35, Harper and Row, New York, 1971.
32. Pinchon, E., "Zapata the Unconquerable," p. 306, Doubleday, New York, 1941.

RECEIVED July 9, 1973.

9

Comparative Analysis of Archaeological Bronzes

W. T. CHASE

Freer Gallery of Art, Smithsonian Institution, Washington, D.C. 20560

While thousands of analyses of archaeological bronzes have been reported in the literature, the basis for comparing them, especially those from different laboratories, is shaky. A round-robin project of chemical analyses was attempted to improve the situation. Two ancient bronze objects were milled to a fine powder, sieved, and mixed to a homogeneous mass. Samples of 500 mg each drawn randomly from this mass were circulated, and results were returned from 21 laboratories. Forty-eight elements were analyzed; some laboratories did only one element, some did as many as 42. The coefficient of variance (or relative standard deviation) ranges from 4% for Cu up to over 200% for some trace elements. The results are tabulated, and methods are suggested to narrow the spread of results in the next run of this program.

The comparative analysis program began at the time "The Freer Chinese Bronzes" (*1, 2*) was being written; R. J. Gettens realized, by studying the literature carefully, that many analyses had been accepted and published without critical appraisal (*3*). The validity and utility of many such analyses seemed open to question. As Gettens said in the earliest draft of the proposal for the comparative analysis program (April 1965) (*4*):

> Most certainly, we all ask ourselves when reading the results of an analysis of an ancient metal object, "Would I have gotten a comparable result if I had sampled and analyzed that object by my own techniques?"

This question may seem simple to answer, but the problem has many aspects which must be considered. Variation in the techniques and methods of sampling is the first aspect. We decided to eliminate this variation in our initial study by taking the samples ourselves (*see* below). Thus, this part of the program concentrates on analytical techniques.

Even among analytical techniques there is great variance. In our study, 10 different analytical methods were used (*see* Table II), and as many different, separate techniques as there are laboratories. In fact, some laboratories employed more than one technique within a given method (—*e.g.*, laboratory 13, which used optical emission spectrometry by both the photoplate and the direct-reading method).

Other variations among laboratories exist—*e.g.*, sample size, number of replicates, elements determined, detection limits, and reporting conventions. To approach the problem of comparable results in the simplest way possible, we took two ancient bronze objects of different composition, reduced half of each object to homogeneous filings, and circulated samples drawn from these filings to various laboratories which had routinely analyzed archaeological bronzes or wanted to participate in the program. The results were tabulated and reported. Our goal was to test the comparability of the results from each laboratory using its usual method under routine conditions.

Now that we have the results, how do they compare? To answer this question fully we have tabulated a detailed, element-by-element examination of the analytical findings below (Table III). In general, the results are fairly good, but discrepancies exist. A careful statistical treatment of the results is very difficult because of several factors (*see* below).

The disparities in the results and the doubts that they raise could be reduced by circulating standards of known composition for bronze analysis and by adopting standard methods for analyzing archaeological bronzes. At present the adoption of standard methods seems impractical because of the differences in equipment and techniques favored by each laboratory. The acquisition and circulation of standard reference materials for bronze analysis are, however, possible and should narrow the spread of results. The program could then be repeated using more samples and a careful statistical design, and sources of variation could be assessed more accurately than can now be done. Of course, the problem of sampling actual objects must also be attacked. The importance of complete reporting (5) when publishing analyses in the field of archaeological chemistry cannot be overstressed.

Procedure

The objects chosen and the procedure used in comminuting them are outlined in our Document No. 2 (*6*) and detailed below.

Sample Preparation. Two bronze objects of limited value as museum display pieces were given to the Freer Gallery by an antique dealer (Simon Kriger, Washington, D.C.). Both pieces were photographed before sectioning. The first object, SC528 or SC-B-1, is a fragment of a Shang Dynasty Chinese bronze *ku*, or beaker, dating from *ca.* 1000 B.C. (Figure

1). It was sectioned in half longitudinally with a diamond saw, and one-half was retained for control and later reference. The other half was reduced to fine turnings with a rotary file in an electric hand drill; before this the outer and inner surfaces were ground to clean metal with another rotary file. The corrosion material was saved but not mixed with the metal sample. No attempt was made to separate iron with a magnet or to remove corrosion from the metal sample by picking it out under the microscope. The sample was passed through a 30-mesh screen, and coarser chips were retained separately. Ninety-nine grams of comminuted sample were produced. This was bottled in glass and shaken to ensure homogeneity. After shaking, part of the sample was rebottled in 0.5-gram (500 mg) lots in glass for distribution.

Figure 1. Top: Shang Dynasty Chinese bronze ku *(wine beaker) in fragmentary condition. Freer Gallery Study Collection No. SC528 or SC-B-1, 13 cm high, before sectioning and comminution. Bottom: after sectioning.*

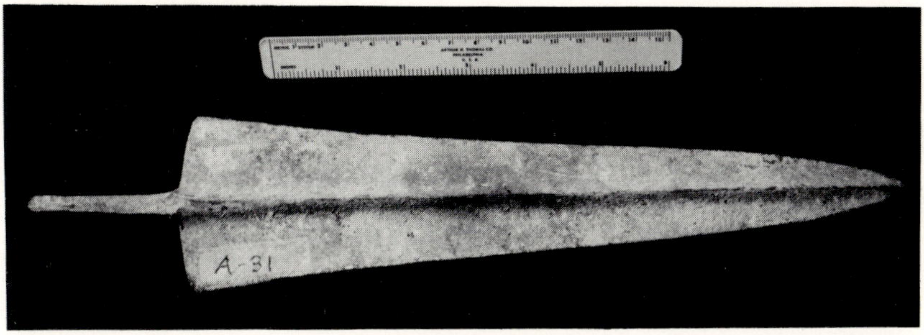

Figure 2. Luristan bronze spearpoint, Freer Gallery Study Collection No. SC615 or SC-B-59 before comminution

The second sample is a Luristan spear point (SC615 or SC-B-59), about 12 inches long, dating from *ca.* 800 B.C. (Figure 2). The corrosion was ground from both sides of half of the blade, from the blade edge to the spine. This half was comminuted as above and sieved to pass 30-mesh. Seventy-nine grams of sample were produced; they were bottled, shaken, and reweighed into 500-mg lots as above.

The shaking, sieving, and repeated shaking were done to make the samples as homogeneous as possible. The bottled 500-mg samples were placed in a box and drawn randomly for circulation. Another sample, sample 3, consisting of 500-mg lots of the U.S. National Bureau of Standards S.R.M. 124c (ounce metal), was sent to laboratories whose results particularly interested us. On reflection, we think that sample 3 should have been sent to all laboratories, and toward the end of the program it was sent to any laboratory that requested samples. We tried to minimize the number of laboratories receiving sample 3 at first because our stock of S.R.M. 124c is very low. The use of other standard reference materials, both as controls and for calibration, will be an important feature of the next phase of the program.

When the samples were sent out, a postcard for acknowledging safe receipt was included. We used the dates on these postcards to mark the start of analysis. On the questionnaire (our Document 3, question 7) a space was supplied for the length of time required for analysis. Few laboratories performed their analyses and returned results within the time they indicated on the questionnaire. This is not surprising. In fact, the length of time required was requested so that we would know when to write the laboratory and ask for their results.

Results

Samples were sent to 36 laboratories, and results were received from 21. In all, 48 elements were analyzed. The results form a complicated picture which we have tried to clarify, while giving all the necessary information, by breaking down the results into a series of tables. Table I shows the laboratories which responded with results; these are listed alphabetically by name of the principal reporter (as received from the laboratory). From Table II on, the laboratories are represented by

Table I. Laboratories and Reporters Represented in Report[a]

Principal Reporter	*Address*
H. Barker	The British Museum Research Laboratory, London WC1B 3DG, England
I. V. Bene	The Freer Gallery of Art Technical Laboratory, Washington, D. C. 20560
L. A. Carrara	Massachusetts Institute of Technology, Central Analytical Facility (Room 13-4153) Cambridge, Mass. 02139
V. F. Hanson	Analytical Laboratory, Henry Francis DuPont Winterthur Museum, Winterthur, Del. 19735
A. Hartmann	Württembergisches Landesmuseum Stuttgart, Chemisch-Physikalisches Labor, D7000 Stuttgart 1, Germany
W. W. Harrison	University of Virginia, Charlottesville, Va.
D. Ishii	Laboratory of Analytical Chemistry, Faculty of Engineering, Nagoya University, Furocho, Chikusaku, Nagoya, Japan
B. B. Johnson	Conservation Center, Los Angeles County Museum of Art, 5905 Wilshire Blvd., Los Angeles, Cal. 90036
F. J. Langmyhr	Department of Analytical Chemistry, University of Oslo, Oslo 3, Norway
L. Maes	Institut Royal du Patrimoine Artistique, 1 Parc du Cinquantenaire 1040, Brussels, Belgium
E. J. Maienthal	National Bureau of Standards, Micro-chemical Analysis Section, Room B331 Chemistry, Washington, D. C. 20234
F. Michel	Laboratoire de Recherche des Musées de France, Palais du Louvre, Place du Carrousel, 75 Paris (1er), France
R. M. Organ	Conservation-Analytical Laboratory, United States National Museum, Smithsonian Institution, Washington, D. C. 20560
C. Patterson and E. Bingham	Spectrographic Laboratory, Division of Geological and Planetary Sciences, California Institute of Technology, Pasadena, Cal. 91109
Rehmatullah	Central Archaeological Laboratory, Old Fort, Lahore, West Pakistan
J. Riederer	Doerner—Institut, D8000 München 2, Meiserstrasse 10, Germany

Table I. Continued

Principal Reporter	*Address*
R. C. Rooney	Rooney and Ward Ltd., Blackwater Station Estate, Camberley, Surrey, England
B. A. Thompson	Analytical Chemistry Division, National Bureau of Standards, Washington, D. C. 20234
C. P. Wan and Li Chi	Institute of History and Philology, Academia Sinica, Nan-kang, Taipei, Taiwan, Republic of China
K. Yamasaki	Department of Chemistry, Faculty of Science, Nagoya University, Nagoya 464, Japan
C. M. Yang	Department of Nuclear Engineering, National Tsing-hua University, Hsinchu, Taiwan, Republic of China

[a] Arranged alphabetically by principal investigator.

number only to preserve anonymity and thus ensure cooperation in future studies. Each laboratory was apprised of its number by private communication.

Table II shows the methods used by each laboratory, the sample sizes used, whether the results came back on our report form or not, and any other pertinent remarks by the laboratory about the samples or procedure. Two laboratories (01 and 04) changed their procedure as a result of our comments on their first analyses. After these changes were made, each laboratory submitted new results. In both cases we retained the old and new results and used the new results for the statistics below. In laboratory 01 the material was given to an analyst relatively inexperienced in the technique on the first run. In laboratory 04, adequate standards for instrument calibration were not available for the first run. No other laboratories had a second chance at these samples, but we hope that when this program is repeated, this sort of interaction between the central data collection facility and the various laboratories will be part of the procedure.

The methods given are presented in Table II exactly as they were reported by the laboratories (as nearly as possible). The detail in which methods are reported varies from elaborate to scanty. No literature references were given by any laboratory. In one case (laboratory 02) results came back on a form marked "report of chemical analysis," and we had to go back to the questionnaire to find that wet chemistry was the method used. Abbreviations are used to reduce space in the method column.

Sample sizes reflect the amount of sample for a single determination (*see* note, Table II) and not the amount of sample consumed by the laboratory.

The column labeled report form used is interesting. Of 23 reports (01 and 04 each gave two), only six used the report form provided. This is why so few sample sizes are listed. While the report form itself (our Document 4) could have been improved and while it did not have space for much information, each laboratory should have filled it in and then provided us with details on additional sheets. In fact, results were much more difficult to compile because so many reporting forms were used; some reports were even handwritten on slips of paper.

The column labeled other remarks is also interesting; many laboratories noted corrosion and inhomogeneity in the samples, especially sample 1. Some laboratories selected sound metal for the analysis.

Table II. Laboratories,

Lab. No.	*Method*
01(old)	ES—emission spectrography
	POL—Davis differential cathode ray polarography
	AA—atomic absorption spectroscopy
01(new)	POL—cathode ray polarography using Davis differential cathode ray polarograph, made by Southern Analytical
	AA—atomic absorption on a Hilger & Watts Atomspek AA spectrophotometer
02	ES—emission spectrography, samples taken as received and arced, qualitative (visual estimate)
	ES—samples compared with NBS standards and SQ powder visually, semiquantitative
	ES—values represent wt %—samples compared with NBS standards by densitometer measurement (averages of 2–4 analyses), semiquantitative
	WET—wet chemistry
04(old)	XRF—non-dispersive x-ray fluorescence spectrometry, Kevex
04(new)	XRF—same
05	ES—optical emission spectrometry, using Cu as an internal standard

Others apparently did not inspect the sample microscopically. The fleck of gold that was discovered in sample 2 by laboratory 31 is interesting (and difficult for us to explain).

Table II shows, in some detail, the complexity and variability of the population involved in this study (laboratories analyzing archaeological materials) and of the reporting procedures. Even the elements analyzed by each laboratory are not completely clear. For example, laboratory 02 reported Cr in trace amounts (0.0001–0.001%) in samples 1 and 2 but gave no report for sample 3 (the space is blank). Thus, we assume that they did not look for Cr in sample 3 although it seems equally likely that they looked for it and it was not present. The proper use of the term "not detected," the consistent reporting of detection limits (done by few laboratories), and the choice of elements to be determined in archaeological bronzes are all open questions.

Methods, and Sample Sizes[a]

Sample Size	*Report Form Used*	*Other Remarks from Laboratory*
NG[b]	no	All samples corroded, especially No. 1
NG		
NG		
ca. 10 mg	no	Sample 1 corroded, thus analyses total <100%
ca. 10 mg		
NG	no	
NG		
NG		
NG		
NG	no	Mentions standards problems
NG	no	
3–4 mg	yes	

Table II.

Lab. No.	*Method*
06	POL—polarography
	NA—neutron activation sample irr. for 20 min. (JR-3, 2×10^{13}N/cm^2/sec), 3 day cooling, diss. in hot HNO_3, add 2 mg. each Ag^+, Au^{2+}, add HCl, heat, add H_2O, heat, cool, filter. Solution extracted with EtAc and counted for ^{198}Au (.411 Mev). AgCl ppt. counted ^{110m}Au (.656 Mev). Gamma-ray counter, NaI (T1) detector, 3 inch ×3 inch.
08	MS—spark source mass spectrometry. For Cu + Sn 100 mg diss. in 5 ml HNO_3/HCl, + soln. made up to 100 ml with dist H_2O; 5 ml of soln. added to 600 mg pure carbon powder. Y used as internal standard. EtOH added and powder dried, pressed, and used as electrode. (Procedure ensures value based on 100-mg sample although only 0.1 mg consumed in spark). Lead content too low to be determined by this method. Diss. of sample 1 in HNO_3 precipitated metastannic acid and indicated a high-tin bronze. Diss. of 2 gave flocculent black precipitate on surface, indicating C or S present.
	MS—solid bronze electrodes made from drillings. Tin content from slurry technique (above) used for internal standard. Rel. sensitivity factors obtained from NBS standards C1115, C1116, C1117, 124, and CA4.
09	EL—electrolysis (wet chemical)
	AA—atomic absorption
	ES—optical emission spectrography
11	XRF—x-ray fluorescence (wavelength-dispersive)
12	POL—polarography, Cu only; sample mixed, diss. in 20 ml dil. HCl + 1 ml HCO_3. Evap. to dryness several times with perchloric acid, hydrobromic acid, and bromine to remove tin. Residues diss. in 10 ml HCl, 20 ml pyridine added, solns dil. to 1 liter. Stds. prep. in same manner from NBS SRM 45d.
13	ES—ultraviolet emission spectrometry with photoplate detection, prism instrument
	ES—ultraviolet emission spectrometry, direct reading type

Continued

Sample Size	*Report Form Used*	*Other Remarks from Laboratory*
10 mg	yes	
10 mg		
100 mg	no	Sample 3 found to be NBS 124 used as standard by lab and excluded from study. Samples examined by stereomicroscope before analysis; both corroded; No. 2 contained large particle of clay. All foreign matter removed from portion for analysis.
5 mg		Accuracy of analysis: (a) major elements by carbon slurry, ±4% (b) trace elements for which standards were available ±10% (c) other trace elements, ±50%
100 mg	yes	Total for sample 1, 97.3%; sample corroded; high Si and Al indicate silicates.
100 mg		
5–10 mg		
70–90 mg	no	
100 mg	no	
NG	no	Sample 3 sent, but not analyzed because of technical difficulty.
NG		

Table II.

Lab. No.	*Method*
15	XRF—(wavelength-dispersive) x-ray spectrometry
	ES—emission spectrometry, semiquantitative
16	WET—wet chemistry
17	AA—atomic absorption
19	NA—neutron activation; for short-lived isotopes (Cu, In, Cl) samples were irradiated in a reactor with thermal neutron flux of 6×10^{13} neutrons/cm^2/sec for 2 secs and counted with a large volume Ge(Li) detector. 3 samples of 10 mg each were used. For longer lived isotopes (Cu, Sn, Co, Ag, Au, and Sb) samples were irradiated for 1 hr, cooled for 7 days, diss. in HNO_3-HF and made up to standard volume for counting. Pure metal foil standards were used for Cu, Sn, Co, Au; pure NH_4Cl was used for Cl and pure Sb_3 for Sb. In calculated from intensity of ^{116m}In, 1.26 Mev γ relative to ^{64}Cu 1.34 Mev γ.
	PA—photon activation, using a linear accelerator
20	AA—atomic absorption; solutions used are 1/1000 dilutions for all the elements.
21	WET—Sn weighed as SnO_2.
	COL—Pb pptd. as $PbSO_4$, diss. in an ammonium acetate soln., and titrated by EDTA with XO as the indicator. Cu titrated by EDTA with PAN as the indicator.
22	WET—sample dissolved in HNO_3, Sn filtered off as metastannic acid and weighed as stannic oxide. Cu + Pb det. simultaneously by electrolysis. Fe pptd. as ferric hydroxide and weighed as ferric oxide. Zn pptd. with diammonium phosphate and weighed as zinc pyrophosphate.
24	MS—mass spectrometry, using electrodes pressed from the samples; spark source MS, against NBS SRM63.
31	WET—Sn; sample diss. in HNO_3 + metastannic acid pptd. SnO_2 weighed; SnI_4 then volatilized using NH_4I, residue analyzed + SnO_2 ppt. weight corrected accordingly.

Continued

Sample Size	*Report Form Used*	*Other Remarks from Laboratory*
ca. 5 mg	yes	
NG		Figures valid within a range of 50% of the reported value, *i.e.*, a concentration reported as .05% might range between .025% and .075%.
200 mg	no	
20 mg	yes	Sometimes difficult to determine tin and silver which may precipitate.
10 mg	no	Material appears inhomogeneous by microscopy.
100 mg	yes	Inhomogeneity suspected; this causes the relatively low precision for the analyses.
ca. 150 mg	no	Insoluble matter not determined in sample 1. In later communication, possible Zn interference in Cu detn. in sample 3 mentioned.
ca. 50 mg	no	Corrosion and foreign products separated microscopically from samples for analysis.
NG	no	
NG	no	

Table II.

Lab. No.	*Method*
	ES—20 mg of sample diss. in HNO_3, evap. to dryness and baked @ 300°C in muffle. Powder mixed with buffers and matrix and burned in dc arc against standards. Error in values *ca.* ±30%.
33	WET—volumetric
	AA—atomic absorption spectrophotometry precision and accuracy estimates in relative %.
34	NA—neutron activation; samples bombarded 3 hrs in thermal neutron flux of 5×10^{12} N/cm²/sec. 10 cc coaxial Ge(Li) detector used. Spectra analyzed by computer program. Totals <100% because short half-life elements (esp. Pb) neglected.

[a] Methods are given with details from the various reporters. Sample size column tabulates sample weight for each determination—*e.g.*, for lab 22, two duplicate deter-

Table III shows the results on the various elements reported. The elements are arranged alphabetically by symbol. Cases where we have extrapolated or averaged data are so marked (—*e.g.*, the averages of results for laboratory 33 in copper). Table III is the most important table here and contains all the information necessary to compile Tables IV, V, and VI. The latter are included to facilitate data comparison and interpretations. Table III is the true result of this phase of the study; although we interpret it in some detail below, this is not absolutely necessary. A superficial perusal of Table III reveals discrepancies which must be caused by systematic errors in the various laboratories. A repeat of the comparative analysis program, with more uniform samples, coupled with circulation of known standards, should reduce this variation.

Table III. Results Reported for Various Elements

Laboratory	*Method*[a]	*Sample 1*	*Sample 2*	*Sample 3*
			Ag (Silver)	
01(old)	ES	present	present	absent
	AA	.018	.089	.021
01(new)	AA	.019	.133	.003
		.017	.132	.003
		Av. .018	Av. .133	Av. .003
02	ES	VFT[c]	T[c]	VFT[c]
		VFT-FT[c]	T[c]	VFT[c]
		.02	.07	.003
04(old)	XRF	.20	.32	.10
04(new)	XRF	.05	.10	.02
05	ES	.020	.14	

Continued

Sample Size	*Report Form Used*	*Other Remarks from Laboratory*
20 mg		Samples inhomogeneous and contain grit from corrosion and casting molds. Can cause large errors—*e.g.*, we found no Au spectrographically in No. 2, yet a fairly large fleck of Au was observed in the ppt. from the HNO_3 diss. of a large quantity of that standard. Grit and sand residues would seriously increase wet Sn values if not corrected.
100 mg 500 mg 50 mg	no no	Atomic absorption used for all elements determined except Cu; although we do not believe that this is the optimum method for some of the elements, we used it to allow us to report a maximum number of those metals; another method would have called for separate samples of .5 g.

minations were carried out on sample 1 with samples of 53.4 mg and 53.9 mg. Cu, Sn, Pb, Fe, and Zn were determined on each.
[b] NG = Not given.

Table III. Continued

Laboratory	*Method*[a]	*Sample 1*	*Sample 2*	*Sample 3*
			Ag (Silver) *Continued*	
06	NA	.020 3*r*	.17 3*r*	
08	MS	.0112	.09	
		.0133	.12	
		Av. .012 ±10%	Av. .1 ±10%	
09	ES	.01 2*r*	.05 −0.1 2*r*	
15	ES	.04	.3	.003
17	AA	.02	.17	
19	NA	.022	.15	
		.016	.14	
		.019	.20	
		Av. .019	Av. .16	
24	MS	.027	.185	
31	ES	.04	.2	
33	AA	.017	.017	
34	NA	.032	.169	0
			Al (Aluminum)	
02	ES	T[c]	VFT[c]	VFT[c]
		FT[c]		
08	MS	.5	.003	
		.4	.003	
		Av. .45 ±50%	Av. .003 ±50%	
09	ES	.026 $\sigma = 0.034\%$ 8*r*	<.002 4*r*	
13	ES[b]	≤.01	<.01	
	ES	≤.01	<.01	
		Av. <.01	Av. <.01	
15	ES	.17	<.0005	<.0005
24	MS	.0035	.0040	
		.0428	.0006	
		Av. .0231 (IVB) [c]	Av. .0023	
33	AA	<.02	<.02	

Table III. Continued

Laboratory	*Method*[a]		*Sample 1*		*Sample 2*	*Sample 3*
					As (Arsenic)	
01(old)	ES		FT[c]		FT[c]	absent
02	ES		FT–T[c]		T–L[c]	—
			FT[c]		T[c]	—
04(old)	XRF		.03		.01	0
04(new)	XRF		0		.06	0
05	ES		.09		.14	
08	MS		.0079		.003	
			.0096		.005	
		Av.	.009 ±10%	Av.	..004 ±10%	
09	ES		~.05 2*r*		.14 2*r*	
13	ES[b]		.10		.20	
			.08		.18	
			.06		.13	
		Av.	.08	Av.	.17	
	ES		~.05		.16	
		Av.	.05 — .08		.16	
				Av.	.16	
				Av.	.165	
15	ES		.04		.12	.02
24	MS		.041		.421	
			.058		.271	
		Av.	.0495 (IVB)[c]	Av.	.346	
31	ES		.02		.1	
34	NA		.0092		.4669	0
					Au (Gold)	
04(new)	XRF		.4		0	0
05	ES		not detected		not detected	
06	NA		.0004 3*r*		.0003 3*r*	
15	ES		<.001		.003	<.001
19	NA		.00036		.0052	
			.00036		.0015	
			.00034		.0039	
		Av.	.00035	Av.	.0003	
31	ES		<.0004		<.0004	
	ES		inhomog. sample		(~.0X) inhomog. sample	
34	NA		.000088		.00025	0
					B (Boron)	
15	ES		.003		.001	.001
					Ba (Barium)	
08	MS				.0006	
					.0003	
				Av.	.0004 ±10%	
15	ES		not detected		not detected	not detected
					Be (Beryllium)	
02	ES		VVFT[c]		absent?	absent?
15	ES		<.0005		<.0005	<.0005
					Bi (Bismuth)	
02	ES		FT[c]		VFT–FT[c]	VFT[c]
04(new)	XRF		.4		0	0
05	ES		.008		.006	
08	MS		.0015		.0025	
			.0015		.002	
		Av.	.0015 ±10%	Av.	.002 ±10%	
09	ES		.011 σ=0.003 4*r*		.0017 σ=0.0002 5*r*	
15	ES		<.001		<.001	<.001
24	MS		.0043		.0039	
31	ES		.005		.001	
33	AA		.010		.003	
					Br (Bromine)	
24	MS		.0038		not detected	
					C (Carbon)	
08	MS		present		present	

Table III. Continued

Laboratory	Method[a]	Sample 1	Sample 2	Sample 3
			Ca (Calcium)	
02	ES	FT–T[c]	FT[c]	VFT[c]
08	MS	.24		
		.5		
		Av. .37 (±50%)		
15	ES	.007	.007	.007
24	MS	.081	INT[c]	
			Cd (Cadmium)	
04(old)	XRF	.08	0	0
15	ES	not detected	not detected	not detected
			Cl (Chlorine)	
19	NA	.80	<.05	
		.34		
		.51		
		Av. .5		
24	MS	.030	.0012	
			Co (Cobalt)	
02	ES	VFT[c]	FT[c]	VFT–FT[c]
		<.01	<.01	<.01
04(old)	XRF	0	.02	0
04(new)	XRF	.4	0	0
05	ES	not detected	not detected	
09	ES	<.001 2*r*	.004 2*r*	
15	ES	<.003	<.003	<.003
17	AA	<.01	<.01	
19	NA	.0004	.00231	
		.0005	.00246	
		.0006	.00250	
		Av. .0005	Av. .0024	
20	AA	not detected	not detected	
24	MS	INT[c]	INT[c]	
			Cr (Chromium)	
02	ES	VFT[c]	VFT[c]	
15	ES	<.001	<.001	<.001
20	AA	not detected	not detected	
24	MS	.048	.0096	
			Cs (Cesium)	
15	ES	not detected	not detected	not detected
			Cu (Copper)	
01(old)	ES	present	present	present
	POL	76.2	90.2	81.2
01(new)	POL	81	94	83.5
			96	
			Av. 95	
	AA	79.6	94.1	82.5
		79.6	93.9	82.1
		82	97	85.0
		80.5	96.5	85.5
		Av. 80.3	Av. 95.4	Av. 83.8
02	ES	H[c]	H[c]	H[c]
		H[c]	H[c]	H[c]
	Wet	78.0	89.6	80.9
04(old)	XRF	86.35	94.7	76.0
04(new)	XRF	82.8	95.6	82.30
05	ES	78.47 (WTC-D)[c]	95.47 (WTC-D)[c]	
06	POL	75.5 ±2.4(3σ,7*r*)	90.2 ±3.0(3σ,7*r*)	
08	MS	81.2	97.2	
		82.5	97.4	
		Av. 81.85 ±4%	95.9	
			94.7	
			Av. 96.25 ±4%	
0	EL	81.36	95.99	
		81.31	96.33	
		Av. 81.34 σ=0.1%	Av. 96.16 σ=0.1%	
11	XRF	81.7	96.5	
		83.5	96.6	
		83.7	95.6	
		82.1	95.6	
		80.9	97.2	
		Av. 82.4 σ=1.2%	Av. 96.3 σ=0.7%	

Table III. Continued

Laboratory	*Method*[a]	*Sample 1*	*Sample 2*	*Sample 3*
			Cu (Copper) *Continued*	
12	POL	81.06	95.61	
		81.11	95.56	
		81.10	Av. 95.59	
		Av. 81.09		
13	ES	84.115 (IVB-D)[c]	95.885 (IVB-D)[c]	
15	XRF	80.7	94.5	81.1
		81.3	91.6	81.3
		83.0	91.8	84.6
		Av. 81.7 ±2.8 C[c]	92.9	Av. 83.6 ±4.9 C[c]
			Av. 92.9 ±3.0 C[c]	
16	Wet	83.05	97.46	
17	AA	83.00	95.00	
19	NA	72.4	89.2	
		78.0	95.7	
		78.8	90.7	
		76.2	91.6	
		74.6	89.7	
		72.2	84.6	
		Av. 75.4	Av. 90.2	
20	AA	80.0	92.50	
		80.0	92.50	
		Av. 80.0	93.75	
			93.95	
			Av. 93.18 $\sigma = \pm 0.784\%$	
21	Wet	83.50	95.55	90.24
22	Wet	82.22	95.48	
		82.56	95.48	
		Av. 82.4	Av. 95.5	
24	MS	Cu+Sn 99.17 (IVB-D)[c]	Cu+Sn 98.80 (IVB-D)[c]	
31	ES	84.75 (IVB-E)[c]	96.05 (IVB-E)[c]	
33	Wet	81.0	95.4	
		81.9	96.0	
		Av. 81.45 (IVB)[c]	Av. 95.7 (IVB)[c]	
34	NA	88.38 ±0.001%	80.55 ±.001%	51.34 ±.001% (IVB)[c]
			Fe (Iron)	
01(old)	ES	T[c]	present	present
01(old)	AA	.05	.21	.13
01(new)	AA	.04	.28	.13
		.05	.27	.13
		Av. .05	Av. .28	Av. .13
02	ES	FT[c]	L[c]	T[c]
		FT[c]	L[c]	T–L[c]
		<.05	.26	.11
	Wet	.08	.22	.06
04(old)	XRF	.19	.24	.24
04(new)	XRF	.4	0	0
05	ES	not studied	not studied	
06	POL	<.1 7*r*	<.1 7*r*	
08	MS	.276	.47	
		.247	.49	
		Av. .26 ±10%	Av. .48 ±10%	
09	ES	.047 $\sigma = 0.014\%$ 4*r*	.19 $\sigma = 0.04\%$ 4*r*	
13	ES[b]	.06	.25	
		.05	.24	
		.01	.20	
		Av. .04	Av. .23	
	ES	~.05	.20	
		Av. .04 — .05	.14	
			Av. .17	
			Av. .20	
15	ES	.08	.25	.12
16	Wet	T[c]	.01	
17	AA	.03	.22	
19	PA	<.1	<.1	
20	AA	.035	.365	
		.040	.280	
		Av. .038	.358	
			.262	
			Av. .316 $\sigma = \pm 0.053\%$	
22	Wet	.66	.51	
		.26	.51	
		Av. .5	Av. .5	
24	MS	.023	.171	
		.087	.228	
		Av. .055 (IVB)[c]	Av. .1995	
31	ES	.07	.2	
33	AA	.054	.27	

Table III. Continued

Laboratory	*Method*[a]	*Sample 1*	*Sample 2*	*Sample 3*
			Ga (Gallium)	
02	ES		VFT–FT[c]	
15	ES	<.001	<.001	<.001
			Ge (Germanium)	
15	ES	not detected	not detected	not detected
			Hg (Mercury)	
09	ES	∼.01 4*r*	.03 2*r*	
15	ES	not detected	not detected	not detected
			In (Indium)	
04(new)	XRF	.3	.45	.12
19	NA	.0020	.00014	
		.0013	.00012	
		.0013	.00012	
		Av. .0015	Av. .00013	
			K (Potassium)	
08	MS	.06		
		.08		
		Av. .07 (±50%)		
15	ES	not detected	not detected	not detected
24	MS	.012	INT[c]	
			Li (Lithium)	
15	ES	not detected	not detected	not detected
			Mg (Magnesium)	
01(old)	ES	T[c]	absent	absent
02	ES	VFT[c]		VVFT[c]
		VFT[c]	VFT[c]	VFT[c]
08	MS	.05	.001	
		.1	.001	
		Av. .075 ±50%	Av. .001 ±50%	
15	ES	.04	.0005	.0005
24	MS	.0068	.0005	
			Mn (Manganese)	
02	ES	VFT[c]		VFT–T[c]
		VFT[c]	VFT[c]	<.003
		<.003	<.003	
04(old)	XRF	.04	.021	.13
04(new)	XRF	.4	0	0
08	MS	.0024		
		.0018		
		Av. .002 ±10%		
11	XRF	.0024	.0034	
		.0028	.0031	
		.0029	.0028	
		.0027	.0027	
		.0027	.0031	
		Av. <.01	Av. <.01	
15	ES	.004	.001	<.001
20	AA	not detected	not detected	
24	MS	.0008	.0005	
33	AA	.0004	.0007	
			Mo (Molybdenum)	
15	ES	not detected	not detected	not detected
			Na (Sodium)	
08	MS	.17	.001	
		.11	.001	
		Av. .14 ±50%	Av. .001 ±50%	
15	ES	not detected	not detected	not detected
24	MS	.020	.0011	
			Nb (Niobium)	
15	ES	not detected	not detected	not detected

Table III. Continued

Laboratory	Methods	Sample 1	Sample 2	Sample 3
			Ni (Nickel)	
02	ES	VFT–FT[c]	L[c]	L–M[c]
		VFT[c]	FT[c]	L[c]
				.67
	Wet	not detected	.05	.65
04(old)	XRF	0	0	.64
04(new)	XRF	0	0	0
05	ES	.006	.020	
09	ES	.0032 $\sigma=0.0002\%$ 4r	.014 $\sigma=0.001\%$ 4r	
11	XRF	.035	.041	
		.025	.032	
		.029	.039	
		.027	.039	
		.028	.033	
		Av. .030 $\sigma=0.01\%$	Av. .04 $\sigma=0.01\%$	
13	ES[b]	<.01	~.01	
	ES	<.01	~.01	
		Av. <.01	~.01	
15	ES	.003	.03	.8
16	Wet	T[c]	T[c]	
17	AA	<.01	.01	
19	PA	<.1	<.1	
20	AA	.011	.017	
		.017	.017	
		Av. .014	.023	
			.017	
			Av. .019 $\sigma=\pm0.003\%$	
24	MS	INT[c]	INT[c]	
31	ES	.005	.009	
33	AA	.013	.015	
			O (Oxygen)	
08	MS	present	present	
			P (Phosphorus)	
08	MS	.003	.0014	
		.003	.0008	
		Av. .003 ±10%	Av. .001 ±10%	
11	XRF	<.001	<.001	
15	ES	not detected	not detected	not detected
24	MS	INT[c]	INT[c]	
			Pb (Lead)	
01(old)	ES	present	T[c]	present
	POL	.53	~.02	3.52
01(new)	POL	.85	.15	4.11
		.73	.12	5.27
		Av. .8	.14	4.3
			Av. .14	Av. 4.42
	AA	.81	.11	4.2
		.82	.10	4.4
		.99	.13	4.6
		.98	.13	4.4
		Av. .9	Av. .12	Av. 4.4
02	ES	T–L[c]	T[c]	M[c]
		L[c]	T[c]	M[c]
		.80	.15	4.9
	Wet	.80	.12	4.7
04(old)	XRF	.7	.090	10.0
04(new)	XRF	1.2	0	4.94
05	ES	1.4	.12	
06	POL	1.20 ±0.03(3σ,7r)	.16 ±0.02(3σ,7r)	
08	MS	.008	.09	
		.114	.08	
		Av. .1 ±10%	Av. .085 ±10%	
09	AA	.90	.19	
		.95	.26	
		.92 $\sigma=0.1\%$ 2r	.23 $\sigma=0.03\%$ 2r	
11	XRF	1.04	.29	
		.94	.27	
		.95	.33	
		.97	.28	
		.96	.31	
		Av. .97 $\sigma=0.04\%$	Av. .30 $\sigma=0.03\%$	
13	ES[b]	.95	.26	
		.92	.20	
		.88	.20	
		Av. .91	Av. .23	

Table III. Continued

Laboratory	*Methods*	*Sample 1*	*Sample 2*	*Sample 3*
			Pb (Lead) *Continued*	
13	ES	1.02	.22	
		.94	.16	
		Av. .98	Av. .19	
		Av. .945	Av. .21	
15	XRF	.56		4.3
		.58		4.2
		.59		4.3
		Av. .58 ±0.04 C[c]	≯.17	4.3 ±0.1 C[c]
	ES		.14	
16	Wet	.823	T[c]	
17	AA	.86	.10	
19	PA	1.2	.15	
		1.2	.16	
		1.2	.14	
		Av. 1.2 (IVB)[c]	Av. .15 (IVB)[c]	
20	AA	not determined	not determined	
21	Wet	1.52	.26	4.38
22	Wet	.97	.16	
		.96	.16	
		Av. 1.0	Av. .2	
24	MS	.253	.166	
		.430	.201	
		Av. .3415 (IVB)[c]	Av. .184	
31	ES	.7	.1	
33	AA	.82	.13	
			Pt (Platinum)	
15	ES	not detected	not detected	not detected
			Rb (Rubidium)	
15	ES	not detected	not detected	not deteected
			S (Sulfur)	
08	MS	.005	.5	
		.004	.54	
		Av. .0045 ±50%	Av. .52 ±50%	
24	MS	.016	.082	
		.024	.066	
		Av. .020 (IVB)[c]	Av. .074	
			Sb (Antimony)	
02	ES	VFT[c]	T[c]	T[c]
			T[c]	T[c]
				.29
04(old)	XRF	.35	.09	.75
04(new)	XRF	.1	.5	.24
05	ES	trace	.10	
		<.02		
08	MS	.0019	.0242	
		.0023	.0245	
		Av. .002 ±10%	Av. .024 ±10%	
09	ES	.008 $\sigma = 0.003\%$ 4r	.057 $\sigma = 0.008\%$ 4r	
13	ES[b]	≤.01	.10	
		≤.01	.10	
		Av. ≤.01	.07	
			Av. .09	
			~.05	
			Av. .05–.09	
15	ES	.01	.05	.25
17	AA	.08	.12	
19	NA	.0075	.082	
		.0083	.077	
		.0079	.078	
		Av. .008	Av. .079	
24	MS	.0053	.188	
		.0077	.143	
		Av. .0065	Av. .166	
31	ES	.004	.03	
33	AA	.019	.12	
34	NA	.0079	.047	.232
			Sc (Scandium)	
15	ES	not detected	not detected	not detected

Table III. Continued

Laboratory	*Methods*	*Sample 1*	*Sample 2*	*Sample 3*
			Se (Selenium)	
08	MS		.002	
			.001	
			Av. .0015 ±10%	
15	ES		not detected	
24	MS	.0043	not detected	
			Si (Silicon)	
01(old)	ES	present	absent	absent
02	ES	L[c]	VFT[c]	VFT[c]
		FT–T[c]	VFT[c]	VFT[c]
		.13	.006	
	Wet	.24	not determined	not determined
08	MS	1.6	.03	
		1.5	.02	
		Av. 1.55 ±50%	Av. .025 ±50%	
09	ES	.1 σ=0.06% 4r	≤.001% 4r	
15	ES	.4	.001	.0005
24	MS	.090	.022	
			Sn (Tin)	
01(old)	ES	present	present	present
	POL	5.33	2.37	2.03
01(new)	POL	14.4	2.9	4.6
			3.2	
			Av. 3.1	
	AA	13.4	2.94	4.55
		13.4	2.94	4.58
		14.4	3.1	4.4
		14.2	3.0	4.2
		Av. 13.85	3.1	Av. 4.45
			Av. 2.97	
02	ES	M[c]	M[c]	M[c]
		M[c]	M[c]	M[c]
		7.8	2.9	4.8
	Wet	13.4	2.2	4.4
04(old)	XRF	11.0	4.28	5.3
04(new)	XRF	12.0	2.2	5.20
05	ES	20	4.0	
06	POL	13.6 ±0.3(3σ,7r)	3.4 ±0.1(3σ,7r)	
08	MS	14.4	3.3	
		13.2	2.6	
		Av. 13.8 ±4%	Av. 2.95 ±4%	
09	AA	14.8	3.8	
		14.9	3.4	
		Av. 14.85 σ=0.3% 2r	Av. 3.6 σ=0.3% 2r	
11	XRF	16.8	2.3	
		16.2	2.5	
		16.3	2.0	
		15.8	2.1	
		15.7	2.3	
		Av. 16.20 σ=0.4%	Av. 2.2 σ=0.2%	
13	ES[b]	14.9	3.8	
		14.6	3.3	
		14.6	3.1	
		Av. 14.7	Av. 3.4	
	ES	15.0	3.6	
		14.8	3.4	
		Av. 14.9	Av. 3.5	
		Av. 14.8	Av. 3.45	
15	XRF	13.0	4.1	5.3
		13.1	3.9	5.6
		13.4	4.1	6.1
		Av. 13.2 ±0.5 C[c]	4.5	Av. 5.7 ±0.9 C[c]
			Av. 4.1 ±0.6 C[c]	
16	Wet	14.69	2.17	
17	AA	13.90	3.40	
19	NA	13.8	3.21	
		14.0	3.24	
		14.4	3.08	
		14.6	3.08	
		13.8	Av. 3.15	
		14.1		
		Av. 14.1		
	PA	13	3.2	
		13	3.2	
		12	3.2	
		Av. 12.66	Av. 3.2 (IVB)[c]	
20	AA	18.85	6.50	
		19.77	6.22	

Table III. Continued

Laboratory	*Method*[a]	*Sample 1*	*Sample 2*	*Sample 3*
			Sn (Tin) *Continued*	
20		Av. 19.31	6.67	
			6.45	
			Av. 6.46 $\sigma = \pm 0.193\%$	
21	Wet	13.82	3.58	5.07
22	Wet	14.90	2.08	
		14.62	1.93	
		Av. 14.8	Av. 2.00	
24	MS	29.7	5.94	
		36.0	4.21	
		Av. 32.85 C[c]	Av. 5.08 B[c]	
		Cu+Sn 99.17 (IVB-D)[c]	Sn+Cu 98.80 (IVB-D)[c]	
31	ES	15	3.5	
	Wet	14.4	3.3	
33	AA	14.3	3.40	
34	NA	5.70	1.33	trace ?
			Sr (Strontium)	
02	ES	VFT[c]		
15	ES	not detected	not detected	not detected
			Ti (Titanium)	
02	ES	VFT–FT[c]	VFT–FT[c]	VFT[c]
15	ES	not detected	not detected	not detected
24	MS	.0010	not detected	
			Tl (Tellurium)	
15	ES	not detected	not detected	not detected
			V (Vanadium)	
15	ES	not detected	not detected	not detected
24	MS	.0024	.0002	
			W (Tungsten)	
15	ES	not detected	not detected	not detected
			Zn (Zinc)	
01(old)	ES	absent	absent	present
	POL			5.2
01(new)	AA	.003	<.001	4.96
		.003	<.001	4.97
		Av. .003	Av. <.001	4.9
				4.9
				Av. 4.93
02	ES			M
				M
				5.6
	Wet	not determined	not determined	4.1
04(old)	XRF	0	0	7.2
04(new)	XRF	0	0	4.94
05	ES	not detected	not detected	
06	POL	<.1 7*r*	<.1 7*r*	
08	MS	.0016	.003	
		.0014	.006	
		Av. .0015 ±10%	Av. .0045 ±10%	
11	XRF	<.01	<.01	
13	ES[b]	~0	~0	
	ES	~0	~0	
		Av. ~0	Av. ~0	
15	ES	not detected	not detected	
	XRF			4.6
				4.4
				4.6
				Av. 4.5 ±0.4 C[c]
17	AA	.01	.01	
19	NA	<.04	<.04	
	PA	<.1	<.1	
20	AA	not determined	not determined	
22	Wet	not detected	not detected	
24	MS	.0027	.0013	
		.0096	.0035	
		Av. .00615	Av. .0024	
31	ES	<.005	<.005	
33	AA	.015	.0039	
34	NA	0	0	15.25

Table III. Continued

Laboratory	Methods[a]	Samyle 1	Sample 2	Sample 3
			Zr (Zirconium)	
15	ES	not detected	not detected	not detected

Laboratory	Methods	Totals[d] Sample 1	Sample 2	Sample 3
01(old)	ES, POL, AA	82.13	92.89	92.10
01(new)	POL, AA	95.12	98.90	97.71
02	ES, Wet	92.55	92.27	95.79
04(old)	XRF	98.94	99.77	100.36
04(new)	XRF	99.9	99.5	98.76
05	ES	postulated as 100	postulated as 100	
06	POL, NA	90.32	93.93	
08	MS	98.70	100.47	
09	EL, AA, ES	97.38	100.34	
11	XRF	99.60	98.84	
12	POL	only Cu analyzed	only Cu analyzed	
13	ES, ES[b]	postulated as 100	postulated as 100	
15	XRF, ES	96.28	97.91	99.31
16	Wet	98.56	99.64	
17	AA	97.90	99.03	
19	NA, PA	91.27	93.83	
20	AA	99.36	99.98	
21	Wet, COL	98.84	99.39	99.69
22	Wet	98.7	98.2	
24	MS	postulated as 100	postulated as 100	
31	Wet, ES	postulated as 100	postulated as 100	
33	Wet, AA	96.72	99.68	
34	NA	94.13	82.56	66.82

[a] All results are given in percentage. The methods used are defined in Table II. Results are reported exactly as received, with replicates reported where we have them. We have not attempted to resolve any ambiguities in the data from each laboratory. In two cases samples were re-run (labs. 01 and 04); both the old and new (second run) data are reported.
[b] Direct reading.
[c] Abbreviations:
A—calculation as received from reporting laboratory
B—"major component best determined by other means"
C—confidence limit using the student's t distribution at the 95% confidence level
D—by difference
E—by difference, using wet analysis value for tin
FT—faint trace
H—high
INT—interference
IVB—I. V. Bene (Freer) calculation from data
L—light
M—major constituent
r—replicates
T or TR—trace
WTC—W. T. Chase (Freer) calculation from data
VFT—very faint trace
[d] Most totals calculated at Freer

Interpretation of Results

Table IV summarizes the results on each element in each sample. The major elements are grouped first, followed by minor and trace elements with more than three determinations (for which the statistics have some validity), minor and trace elements with less than three determinations, and elements singly determined. We examine this table in more detail under Comparisons with Similar Studies below.

Tables Va, Vb, and Vc summarize the results for copper which has by far the lowest relative standard deviation of any element in our study. This means that it has been generally the most precisely determined element. Some laboratories, on the other hand, consider that copper, being the major constituent, does not need to be determined; these include laboratories

05, 13, and 31, all of which used optical emission spectrometry primarily to determine trace elements. We calculated the copper percentages for these laboratories by difference, and while these percentages are not included in the statistics, they are remarkably close to the mean value. For the most part the values calculated by difference will be ignored. In describing an object made from copper or a copper alloy, the copper content is a datum of prime importance. How well have the various laboratories been able to determine the copper content? What method seems to be the best?

Tables Va, b, and c attempt to answer these questions. In Table Va percent copper is ranked in increasing order along with the laboratory numbers and the methods. First, are any of the values outliers? Should all the values be taken as representative of the copper content, or are some in error? Assuming that the data are normally distributed (*i.e.*, that they follow the Gaussian or bell-curve distribution) the "*r*" test can be applied to reject anomalous values; this was done, and four values were rejected. Three come from one laboratory, and the fourth came from a laboratory which did not detect or take Zn into account on sample 3. Therefore, their calculation for copper was too high.

Thus we have 16 determinations (without the difference values) from which to calculate the mean in samples 1 and 2, and four in sample 3. Sample 3 (NBS SRM 124c) has a known value of 84.22%, with which we can compare our value (from four laboratories) of 82.6%. Our mean is somewhat low. The median, higher in each case, may be a better estimate of what the copper content actually is in each sample. The standard deviation and relative standard deviation are shown (*see* footnote b, Table IV). The relative standard deviation decreases from sample 1 to sample 2 and from sample 2 to sample 3. This may reflect greater inhomogeneity in sample 1 than sample 2 and in sample 2 than sample 3; it may also simply reflect the fact that we have fewer laboratories reporting on sample 3. Since the relative standard deviation on copper is so low, some of it may indeed arise from the materials themselves. On the other hand, the relative standard deviations for other elements (Table IV) do not show this pattern. In fact, they seem to be more closely related to the amount present, implying that the determinations become less precise as the amount of the element decreases (*viz.*, Sn 1 + 2; Zn 1, 2, + 3; Ag 2 + 3; Si 1 + 2).

While the Youden plot (Figure 3) indicates that there is systematic error in the various laboratories, a test by Spearman's rank correlation coefficient (Table Vb) is ambiguous. The ranking is significant at the 0.05 level but not at the 0.01 level; thus a correlation of the ranks of the laboratories such as we see here would appear less than five times in 100 by chance but more than one time in 100. I prefer to look at these results in terms

Table IV. Summary

Elem.	Sample 1 Mean	Rel.[b] SD, %	No. of Val.[c]	Range[d]	Sample 2 Mean	Rel.[b] SD, %	No. of Val.[c]	Range[d]
Group I—Major Elements								
Cu	80.86	3.1	16	75.4–83.5 (78.47, 84.115, 99.17, 84.75, 88.38)	94.40	2.6	16	89.6–97.46 (95.47, 95.885, 98.80, 96.05, 80.55)
Sn	14.72	14	18	12.0–20 (5.7, 32.85)	2.96	25	18	1.33–4.1 (5.08, 6.46)
Pb	.900	38	18	.1–1.52	.167	38	16	.085–.3(0)
Zn	.00713	77	5	.0015–.015 (3[0], <.005, 2[<.01], <.04)	.0052	64	4	.0024–.010 (3[0], <.001, <.005, <.01, <.04, <.1)
Group II—Minor and Trace Elements $n > 3$								
Ag	.0246	47	14	.01–.05	.142	48	14	.017–.3
As	.0435	70	8	.009–.09(0)	.172	84	9	.004–.4669
Bi	.00663	55	6	.0015–.011 (<.001, .4)	.00293	62	6	.001–.006 (0, <.001)
Fe	.0549	31	10	.030–.080 (2[<.1], .26, .4, .5)	.258	49	13	.010–.5 (0,2[<.1])
Mn	.00198	74	5	.0004–.004 (<.003, .4)	.0013	88	4	.0005–.003 (0, <.003)
Ni	.0106	91	7	.003–.03 (0,2[<.01], <.1)	.0217	64	10	.009–.05 (0, <.1)
Sb	.0230	140	12	.002–.1	.0803	55	11	.024–.166 (T, .5)
Si	.208	70	4	.09–.4 (1.55)	.0110	106	5	.001–.025
Group III—Minor and Trace Elements $n \leq 3$								
Al	.0730	115	3	.0231–.17 (<.01, <.02, .45)	.00265	19	2	.0023–.003 (<.0005, <.002, <.01, <.02)
Au	.000279	60	3	.000088–.0004 (<.0004, <.001, .4)	.000962	141	4	.00025–.003 (0, <.0004)
Ca	.153	126	3	.007–.37 (FT–T)	.007	—	1	(FT,INT)
Cl	.265	125	2	.030–.5	.0012	—	1	(<.05)
Co	.0005	—	1	(<.001, <.003, 2[<.01], .4)	.00320	35	2	.0024–.0040 (0, <.003, 2[<.01])
In	.151	139	2	.0015–.3	.225	141	2	.00013–.45
K	.041	100	2	.012–.07 (not detected)	not detected	—	1	(INT)
Mg	.0406	84	3	.0068–.075 (VFT)	.000667	43	3	.0005–.001 (VFT, absent)
Na	.080	106	2	.020–.14 (not detected)	.00105	6.7	2	.001–.0011 (not detected)
S	.0122	90	2	.0045–.020	.297	106	2	.074–.52
Group IV—Elements with Only One Determination[f]								
B	.003		1	—	.001		1	—
Ba	not detected		1	—	.0004		1	(not detected)
Be	—		1	(VVFT)	<.0005		1	(absent)
Br	.0038		1	—	not detected		1	—
C	present		1	—	present		1	—
Cr	.048		1	(<.001,VFT not detected)	.0096		1	(<.001,VFT not detected)
Ga	<.001		1	—	<.001		1	(VFT–FT)
Hg	.01		1	(not detected)	.03		1	(not detected)
O	present		1	—	present		1	—
P	.003		1	(<.001, not detected, INT)	.001		1	(not detected, INT)

of Results[a]

Sample 3				
Mean	*Rel.[b] SD, %*	*No. of Val.[c]*	*Range[d]*	*NBS Val.[e]*
Group I—Major Elements				
82.6	1.6	4	80.9–83.6 (51.34, 90.24)	84.22
4.97	10.7	5	4.4–5.7(trace)	5.13
4.54	5.9	5	4.3–4.94	4.74
4.62	8.7	4	4.1–4.94 (15.25)	4.93
Group II—Minor and Trace Elements n <3				
.00725	117	4	.003–.020(0)	
.02	—	1	.02(0,0)	
<.001	—	1	(0)	
.103	37	3	.060–.13 (0)	.11
.003	—	1	.003(0, <.001)	
.725	15	2	.65–.8(0)	.60
.253	10	4	.232–.29	.20
.0005	—	1	.0005(VFT)	.002
Group III—Minor and Trace Elements n ≤ 3				
<.0005	—	—	—	
<.001	—	1	(0,0)	
.007	—	1	(VFT)	
—	—	None	—	
<.01	—	1	(0, <.003)	
.12	—	1	—	
not detected	—	1	—	
.0005	—	1	.0005 (absent, VFT)	
not detected	—	1	—	
—	—	None	—	.048
Group IV—Elements with Only One Determination				
.001		1	—	
not detected		1	—	
<.0005		1	(absent)	
present		1	—	
—		None	—	
<.001		1	—	
<.001		1	—	
not detected		1	—	
—		None	—	
not detected		1	—	.024

Table IV.

	Sample 1				Sample 2			
Elem.	*Mean*	*Rel.*[b] *SD, %*	*No. of Val.*[c]	*Range*[d]	*Mean*	*Rel.*[b] *SD, %*	*No. of Val.*[c]	*Range*[d]
Se	.0043		1	—	.0015		1	(2[not detected])
Sr	VFT		1	(not detected)	not detected		2	—
Ti	.0010		1	(VFT–FT, (not detected)	VFT–FT		1	(2[not detected])
V	.0024		1	(not detected)	.002		1	(not detected)

[a] Elements arranged alphabetically within groups except group 1. First group has the four elements commonly occurring above 1% in ancient copper alloy samples which we have analyzed at the Freer; Cu, Sn, Pb, Zn. Next group has minor and trace elements for which more than three determinations have been used to calculate the mean in sample 1 or 2. Next group has minor and trace elements for which three or two determinations have been used to calculate the mean. Last group has elements for which single determinations only are available. The mean represents what the samples actually contain.

[b] Relative standard deviation (or coefficient of variance) is the standard deviation divided by the mean.

$\frac{s}{\bar{x}}$ = relative standard deviation where $\bar{x}$ = the mean and s = the standard deviation.

$$s = \sqrt{\frac{\sum_i (x_i - x)^2}{n-1}}$$ where n = the number of values, x_i = individual value number i, and $\bar{x}$ = the mean.

The relative standard deviation enables us to look at the variability in the results for each element, disregarding the absolute amount.

of the Youden plot, a scheme devised for comparative testing (7). It is not completely valid here since the two materials should be present in the same amounts. The copper contents in sample 1 and 2 do approximate this, however. The results for the various laboratories are graphed on a two-dimensional plot of the amount of copper in sample 1 *vs.* sample 2. The results are not too widely scattered; with three exceptions they fall into the first and third quadrants, and, with these exceptions, fall within an ellipse around the 45° line drawn through the point determined by the mean for samples 1 and 2. Thus, if a laboratory is high on the copper content in sample 1, it is high on the content in sample 2; if it is low on sample 1, it is low on sample 2. The precision (shown by the distance from the 45° line) is quite good in most cases, but systematic errors are clearly present. Much of this error could be eliminated by calibrating the procedures used by the laboratories against standard reference materials chosen and circulated in the next round of our project. The scatter of our results is pointed up by the linear plots (Figure 4, after Dozinel) shown along with one of Dozinel's own plots (8). We certainly show worse scatter than he does.

In Table Vc we attempt to show which method is best. Since there are few results for any given method (a maximum of six in the case of wet chemistry), a simple scheme was chosen. The absolute values of the deviations around the mean were added and then divided by the number of determinations to give an average deviation for the method on each sample. The average deviations on samples 1 and 2 were again averaged to give an average deviation for the methods. Then the methods were ranked. Interestingly, atomic absorption is the best. Mass spec-

Continued

Sample 3

Mean	*Rel.*[b] *SD, %*	*No. of Val.*[c]	*Range*[d]	*NBS Val.*[e]
—		None	—	
not detected		2	—	
VFT		1	(not detected)	
not detected		1	—	

[c] Number of individual values used to calculate the mean and standard deviation.
[d] Values in parentheses are those discarded in making the mean and standard deviation calculations. When in doubt, Massey's criterion (*20*) for testing for extreme mean is used, at the .05 level of significance. Where only one determination has been used, the result is listed in the "means" column, with any other data listed, in parentheses, under the "range." The single determinations are not repeated in the "range" column.
[e] Value on NBS certificate of analysis for SRM 124c ounce metal.
[f] The elements Cd, Cs, Ge, Li, Mo, Nb, Pt, Rb, Sc, Tl, W, and Zr were looked for in all three samples but not detected (one determination each).

trometry is next, but since only one result was used, this has little statistical validity. X-ray fluorescence is third, and wet chemistry is fourth. This indicates poor determinations by some of the wet chemists. Polarography falls close to the bottom of the list.

An analysis of variance was run, but this did not indicate any surprising results. It did show that the difference between laboratories was greater than would be expected by chance. The analysis of variance approach could be facilitated in the future by having the same number of replicates run by each laboratory and by increasing the number of materials analyzed.

Table VI shows the major element composition of the samples as determined by the various laboratories. This table was compiled to emphasize that while there are large discrepancies in the results, most of the laboratories could characterize the samples correctly. Thus, sample 1 is a moderately high tin bronze (Sn *ca.* 15%) with about 1% lead and little iron or zinc. Laboratories that fail on the tin value in this characterization are 01 (old), and 04 (old), possibly 05 with a tin value of 20% (although this is an optical emission spectrographic value and falls in the right range), 24 (which doesn't claim any accuracy for its tin result), and 34. Laboratory 01 (old) also has a low lead value as do 08 and 24. Thus, six of 23 laboratories (or about 25% of the results) fail to characterize the samples correctly while the other 17 characterize this sample as a moderately high tin bronze with a little lead.

Sample 2 is a low-tin bronze (or tin-alloyed copper) with very little lead (less than 0.5%), a little iron, and practically no zinc. Almost all of the labs reported this correctly. The tin results on 04 (old), 05, 15, 20, and 24 are somewhat high. The characterization is at least as good as on sample 1.

In sample 3, which was only sent to certain laboratories, the results may be somewhat clearer. This sample is a standard reference material from the U. S. National Bureau of Standards (SRM 124c). It is an ounce metal with a nominal composition of 85-5-5-5—that is, 85% Cu and 5%

Table Va.

	Sample 1		
	Values in Increasing Order	*Lab.*	*Method*
	75.4	19	NA
	75.5	06	POL
	78.0	02	Wet
	78.47-D[a]	05	ES
	80.0	20	AA
	80.3	01	AA
	81.09	12	POL
	81.34	09	EL
	81.45	33	Wet
	81.7	15	XRF
	81.85	08	MS
	82.4	11	XRF
	82.4	22	Wet
	82.8	04	XRF
	83.0	17	AA
	83.05	16	Wet
	83.5	21	Wet
	84.115-D[a]	13	ES
	84.75-D[a]	31	ES
	88.38[a]	34	NA
Number of values used to compute statistics	16		
Mean	80.8612		
Median	81.58		
Standard deviation	2.51586		
Coefficient of variance (relative standard deviation), %	3.11		

[a] Values not included. Outliers were tested as follows. For copper, each extreme value was tested. The test used to decide if the extreme value were in error was the "*r*" test as applied by Massey (*20*), Dixon (*21*), and Proschan (*22*). The test was applied at the 0.5 level of significance; thus, a true value would be rejected 5% of the time. Formulas used were as follows:

$$r_{22} = \frac{x_3 - x_1}{x_{k-2} - x_1} \qquad r_{10} = \frac{x_2 - x_1}{x_k - x_1}$$

Decision points: $n = 16$, $r_{22} \leq .507$; $n = 6$, $r_{10} \leq .560$; $n = 5$, $r_{10} \leq .642$. For r with each of Sn, Pb, Zn. The actual values as determined by NBS analysts are 84.22% Cu, 5.13% Sn, 4.74% Pb, and 4.93% Zn. Laboratories failing to characterize this sample correctly are 01 (old), 04 (old), 21 (which did not look for zinc), and 34. The result for laboratory 02 is a little low in copper, but the presence of approximately equal parts of Sn, Pb, and Zn are noted. Thus for eight reports we have four that are in error as to sample characterization. Of course, if the earlier reports from

Copper Results

Sample 2			*Sample 3*		
Values in Increasing Order	*Lab.*	*Method*	*Values in Increasing Order*	*Lab.*	*Method*
80.55[a]	34	NA			
89.6	02	Wet			
90.2	06	POL			
90.2	19	NA			
92.9	15	XRF			
93.18	20	AA	51.34[a]	34	NA
95.0	17	AA	80.9	02	Wet
95.25	01	AA	82.3	04	XRF
95.47-D[a]	05	ES	83.6	15	XRF
95.5	22	Wet	83.6	01	POL
95.55	21	Wet	90.24[a]	21	Wet
95.59	12	POL			
95.6	04	XRF			
95.7	33	Wet			
95.885-D[a]	13	ES			
96.05-D[a]	31	ES			
96.16	09	EL			
96.25	08	MS			
96.3	11	XRF			
97.46	16	Wet			
16			4		
94.4025			82.6 (NBS value = 84.22)		
95.52			82.95		
2.44558			1.28842		
2.59			1.56		

these values, the datum being tested is retained as a true value. r_{22} (75.4) = .339; thus, 75.4 retained. r_{22} (80.55) = .614; thus, 80.55 discarded. r_{10} (51.34) = .759; thus, 51.34 discarded. r_{22} (88.38) = .513; thus, 88.38 discarded. r_{22} (97.46) = .166; thus, 97.46 retained. r_{10} (90.24)* = .710; thus, 90.24 discarded [*r_{10} (90.24) was calculated after discarding 51.34].

Table Vb. Applying

Laboratory	01	02	04	06	08	09	11
Rank on sample 1	5	3	12	2	10	7	11
Rank on sample 2	6	1	10	2	13	12	14
Difference	−1	2	2	0	−3	−5	−3
Squared difference	1	4	4	0	9	25	9

Sum of squared differences = 210; pairs correction (two pairs of two

laboratories 01 and 04 are disregarded, we have six reports with two in error; this proportion is not vastly different from the proportions on samples 1 and 2. Our results should be better than this. We cannot discuss in detail the questions of inhomogeneity of samples 1 and 2 affecting the results, but there is no question of inhomogeneity in sample 3—a prepared standard reference material—at least in amounts generally used by these analysts (10 mg or more in most cases). In fact, the lowest sample size, laboratory 15 with *ca.* 5 mg, has yielded values which are quite close to the true values. Before suggesting remedies, let us relate the results of our comparative analysis program to other programs of comparative analysis.

Table Vc. Examination

Method	*Deviations*[b]
Wet	−2.86, .589, 1.54, 2.19, 2.64, .479 −4.802, 1.098, 1.148, 1.298, 1.758, 3.058
NA	−5.46, 7.519 −13.853, −4.202
AA	−.861, 2.139, −.561 −1.222, .598, 1.848
POL	−5.361, .229 −4.202, 1.188
ES	−2.391, 3.25, 3.889 1.068, 1.483, 1.648
XRF	.839, 1.539, 1.939 −1.503, 1.197, 1.897
MS	.989 1.848

[a] Evaluation of method errors: Deviations for each method around the calculated mean were determined, then the absolute values of these deviations were added. The sum of the absolute values of the deviations was divided by the number of values to give an average deviation for each method on each sample, and these averages then averaged again.

[b] First line = sample 1; second line = sample 2.

Spearman's Correlation Test (*23*)

12	15	16	17	19	20	21	22	33
6	9	14	13	1	4	15	11	8
9	3	15	5	2	4	8	7	11
−3	3	−1	8	−1	0	7	4	−3
9	9	1	64	1	0	49	16	9

observations each) = ½ × 2 = 1. $D^2 + T = 211$; this value is significant at the .05 level but not at the .01 level.

Comparison with Other Interlaboratory Studies

Much literature has appeared on interlaboratory comparison studies. Most of this, especially in the statistical field, bears directly on the question of comparison of test methods, or interlaboratory studies of a test method—the question of standardizing methods (*7, 9, 10, 11*). While useful information and concepts can be gleaned from these publications, especially from Youden and Mandel, our problem is more complex. We are trying to evaluate the performance of various laboratories, using various methods, on single materials. The reports which bear directly on this question are for the most part not statistically elegant and gen-

of Copper Results[a]

Σ *Deviation*	n	Σ *Deviation*/n	*Average*[c]
10.298	6	1.716	1.955
13.162	6	2.194	
12.979	2	6.490	7.759
18.055	2	9.028	
3.561	3	1.187	1.205
3.668	3	1.223	
5.590	2	2.795	2.745
5.390	2	2.695	
9.530	3	3.177	2.288
4.199	3	1.399	
4.317	3	1.439	1.486
4.597	3	1.532	
.989	1	.989	1.419
1.848	1	1.848	

[c] Methods ranked in order of average deviation (lowest to highest): AA, MS, XRF, Wet, ES, POL, NA. Of course, one must consider the small number of values on which this ranking is based; MS (the second-best method) is based on only one value. The low place of wet chemistry on this table does bear some consideration.

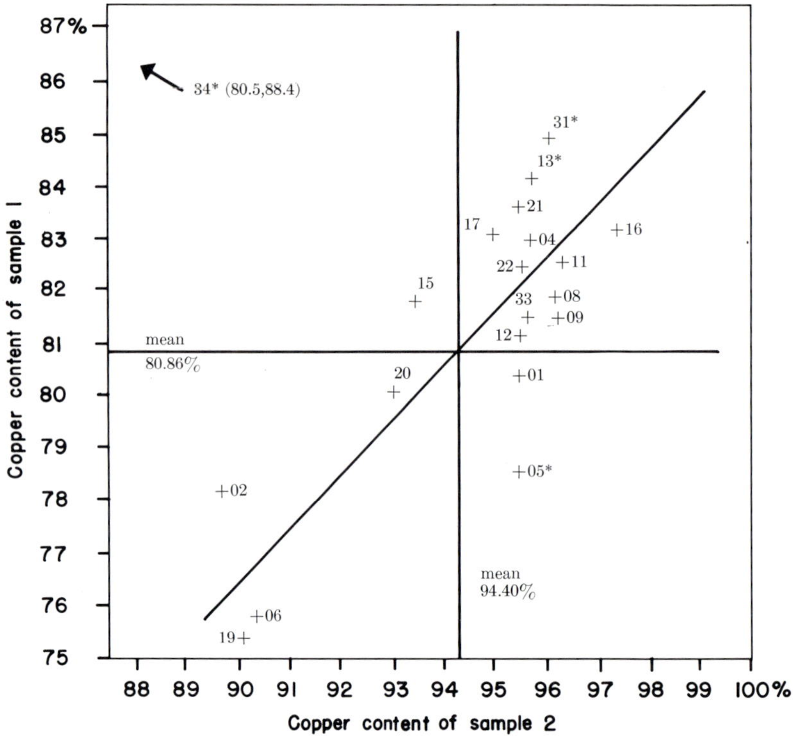

Figure 3. Youden plot (see *Ref. 7*) *of copper content in sample 2* vs. *copper content in sample 1. Each point corresponds to the two "best estimates" of the copper contents from a particular laboratory, and the laboratory number is given next to the point. The indicated means are calculated in Table V. Laboratory results not included in the calculation of the means are marked with an asterisk.* A 45° *line* (X = Y + K) *has been drawn in to indicate the spread of values.*

erally show the present state of affairs, with explanations and suggestions for improvement. The data are treated in many different ways. Let's look first at Morrison's study of the lunar elemental analyses (*15*). Here he has treated the data rationally and concisely; in fact, the form of our Table IV is drawn directly from this paper. The methods he cites (neutron activation, mass spectrometry, x-ray fluorescence, wet chemistry, emission spectrometry, and others) cover the methods in our study. How do our results compare with his? In major elements (for Morrison, Al, O, Si, Fe, Ca, Mn, P, Na, Ti, K, Mg, and Cr), has relative standard deviation ranges from a low of 1.8% (Al; mean value 6.7%, number of laboratories reporting, 11), to a high of 10% (Mg and Cr; means 6.35% and 0.267% respectively, number of laboratories reporting, nine and 13). Excluding the low zinc amounts in samples 1 and 2, our relative standard deviations range from 1.6 to 38%. Our mean relative standard deviation on Cu, Sn, and Pb in all samples and Zn in sample 3 is 14.8%—higher than Morrison's highest relative standard deviation. For the minor and trace category, let us compare two specific elements:

	Our Results			*Morrison's Results*		
	Mean	*Rel. SD, %*	*No.*	*Mean*	*Rel. SD, %*	*No.*
Ni:	0.0106	91	7			
	0.0217	64	10	0.000195	8	11
	0.725	15	2			
Au:	0.000279	60	3			
	0.00962	141	4	3.0×10^{-8}	55	3

In almost all cases our relative standard deviations are higher than Morrison's. This shows that we have a greater spread of results.

Flanagan, working with USGS standard rocks (pieces of the Westerly Granite), shows an 11% relative standard deviation for Fe (in an amount around 1%); much less than our value of 31–49% (*12*). The manganese determinations he reports have a relative standard deviation of 43%, again better than our 74% and 88%. We also have a greater spread of results than his.

Our results on copper do not appear to have a greater spread than those of Schweizer and Schweizer and Friedman for gold and for silver, but relative standard deviations have not been calculated (*17, 18*).

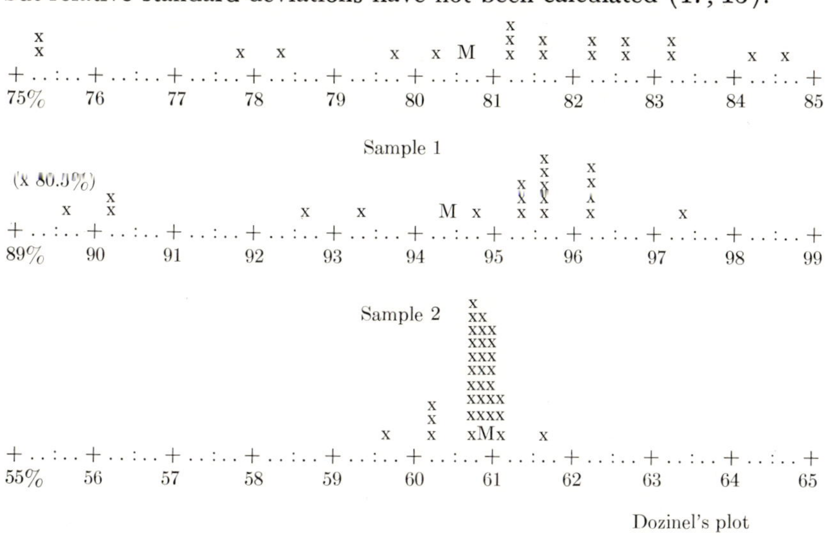

Figure 4. Comparison of copper determinations on our samples 1 and 2 with values from 32 industrial laboratories (Ref 8, p. 268). Each x indicates one analytical result; M = mean value.

Table VI. Summary of Major

Sample 1

Lab	*Cu*	*Sn*	*Pb*	*Fe*	*Zn*	*Total*
01(old)	76.2	5.3	.53	.05	absent	82.1
01(new)	80.3	13.8	.85	.05	.003	95.0
02	78.0	13.4	.80	.08	—	92.3
04(old)	86.4	11.0	.7	.19	0	98.3
04(new)	82.8	12.0	1.2	.4	0	96.4
05	78.5[d]	20	1.4	—	not detected	[c]
06	75.5	13.6	1.20	<.1	<.1	90.5
08	81.8	13.8	.1	.26	.0015	95.9
09	81.3	14.8	.92	.047	—	97.1
11	82.4	16.2	.97	—	<.01	99.6
12	81.1	—	—	—	—	—
13	84.1[d]	14.7	.91	.045	0	99.8
15	81.7	13.2	.58	.08	not detected	95.6
16	83.1	14.7	.82	trace	—	98.6
17	83.0	13.9	.86	.03	.01	97.8
19	75.4	14.1	1.2	<.1	.04	90.8
20	80.0	19.3	—	.038	—	99.3
21	83.5	13.8	1.52	—	—	98.8
22	82.4	14.8	1.0	.5	not determined	98.7
24	—	32.8[a]	.34	.55	.006	[c]
31	84.2[d]	14.4	.7	.07	<.005	99.4
33	81.4	14.3	.82	.054	.015	96.6
34	88.4	5.7	—	—	0	94.1
NBS values						

[a] Major component, best determined by other means,—*i.e.*, the analyst has little confidence in this value.
[b] Sample 3 is standard reference material 124c from the United States National Bureau of Standards. The last line (NBS values) shows the values on the analysis certificate, based on a number of wet determinations for each element using 1-gram samples. The alloy also contains P = .024%, Si = .002%,

Croegaert's study of the results of the Corning interlaboratory analysis of glasses project, organized by Robert Brill, also shows wide variation (*19*). In the major elements, silicon (in their sample A) ranges (calculated as SiO_2) from 63.3 to 70%, or a 6.7% range; our copper in sample 1 shows an 8.1% range. The next most abundant constituent, sodium (calculated as Na_2O) shows a range in their sample A of 2.83%; our tin in sample 1 has a range of 8%. Standard deviations were not calculated for the Corning results. The copper ranges in a study cited by Dozinel can be compared graphically (Figure 4) (*8*). It is clear which is better.

Clearly the procedures of the laboratories involved in this project could be improved. Before suggesting methods of improvement, one obvious point should be stressed. The results of this study do not vitiate the use of data between laboratories; they merely indicate that the data should be used with caution. For example, in the Luristan bronzes (*14*), the arsenic content of arsenical copper artifacts is said to drop from 2–6% to 2% and below after a tin–copper alloy was introduced about 2600 B.C. If another laboratory analyzes a copper alloy object and finds, say 4.9% arsenic (and all other factors agree), the object can be confidently set into the earlier group; if the laboratory finds 0.8% arsenic, the object could be set into the later group. If the arsenic content is 1.95%, the grouping into which it would fall is less unequivocal, and other factors

Elements by Laboratory

Sample 2						Sample 3[b]					
Cu	*Sn*	*Pb*	*Fe*	*Zn*	*Total*	*Cu*	*Sn*	*Pb*	*Fe*	*Zn*	*Total*
90.2	2.4	.02	.21	<.001	92.8	81.2	2.0	3.5	.13	5.2	92.0
95.2	3.0	.13	.28	<.001	98.6	83.6	4.5	4.4	.13	4.9	97.5
89.6	2.2	.12	.22	—	92.1	80.9	4.4	4.7	.06	4.1	94.2
94.7	4.3	.09	.24	0	99.3	76.0	5.3	10.0	.24	7.2	98.7
95.6	2.2	0	0	0	97.8	82.3	5.2	4.9	0	4.9	97.3
95.5[d]	4.0	.12	—	not detected	99.6						
90.2	3.4	.16	<.1	<.1	93.8						
96.2	3.0	.08	.48	.0045	99.8						
96.2	3.6	.23	.19	—	100.2						
96.3	2.2	.30	—	<.01	98.8						
95.6	—	—	—	—	—						
95.9[d]	3.4	.23	.22	0	99.8						
92.9	4.1	≯.17	.25	not detected	97.4	83.6	5.7	4.3	.12	4.5	98.2
97.5	2.2	trace	.01	—	99.7						
95.0	3.4	.10	.22	.01	98.7						
90.2	3.2	.15	<.1	<.04	93.6						
93.2	6.5	—	.32	—	100.0						
95.6	3.6	.26	—	—	99.4	90.2	5.1	4.4	—	—	99.7
95.5	2.0	.2	.5	—	98.2						
—	5.1[a]	.18	.20	.0024	[c]						
96.0[d]	3.3	.1	.2	<.005	99.6						
95.7	3.4	.13	.27	.0039	99.5						
80.6	1.3	—	—	0	81.9	51.3	trace	—	—	15.25	66.6
						84.22	5.13	4.74	.11	4.93	99.02

Sb = .020%, S = .048%, Ni = .60%
[c] Postulated as 100%.
[d] By difference.

(other alloy constituents, date, shape, surface finish, corrosion, etc.) would have to be considered. In other words, there are areas of uncertainty around the results, and these uncertainties must be considered when the results are used.

Of course, if the results on a group of objects are obtained by a single laboratory using a single method under adequate control, the areas of uncertainty can be kept small; in this case, the data can be used to answer more detailed questions than when interlaboratory data are used.

Present Needs in the Program

How can we narrow the spread of the results? What improvements can be made in our procedure? The comparative analysis program should be run again. In our second run, the entire system of reporting should be redesigned. It should be made clear that results will be ignored unless they are reported on the correct form with all pertinent information. All of the information requested by R. M. Organ (5) should be present on the reporting form. Literature references for any methods used (where available) should be given. Limits of detection for each element sought should be given. Figures for assessing the variance or standard deviation of each method used should be reported (a series of duplicate or multiple determinations, a series of determinations on standard materials, etc.).

These can be independent of our circulated samples. Inhomogeneity of of the samples should be carefully controlled and assessed.

Known samples with stated values of the elements of interest should be circulated, along with the unknowns, for calibrating the methods used in the various laboratories. Standards of known composition should also be included with the unknowns. It is possible that "standard archaeological bronzes" should be synthesized for this program and for future circulation. Also, the statistical basis of the program should be reassessed and a new statistical design drawn up so that data collection and analysis can be simplified. Since many archaeologists and museums use the services of commercial laboratories, samples should be circulated to these laboratories for paid analysis.

The next steps in the program are as follows:

(1) A report will be circulated to the various laboratories along with a questionnaire to determine the following:

- (a) Do these materials represent the range of interest of your laboratory?
- (b) What elements, and in what amounts, do you normally search for?
- (c) Would you be willing to participate in another round of the program?
- (d) Do you know of any other laboratory that should participate?
- (e) Are you in agreement with the conclusions in this report?

(2) A new statistical design for the program should be made.

(3) The ranges and compositions of unknowns and standard samples should be decided.

(4) Materials will be procured, and another round of the program will be run.

In the future we should consider the problem of sampling archaeological artifacts in detail. Also, the question of standard analytical methods for various classes of archaeological artifacts should be considered.

This report has made a small beginning in rationalizing the results from different laboratories analyzing archaeological objects of metal. Even if it serves no other purpose, it clearly indicates that these analyses must be performed, reported, and interpreted with care if the results are to be valid in archaeological studies.

Literature Cited

1. Pope, J. A., Gettens, R. J., Cahill, J., Barnard, N., "The Freer Chinese Bronzes, Volume I, Catalogue," Freer Gallery of Art, Oriental Studies, No. 7, Washington, D.C., 1967.
2. Gettens, R. J., "The Freer Chinese Bronzes, Volume II, Technical Studies," Freer Gallery of Art, Oriental Studies, No. 7, Washington, D.C., 1969.

3. Barnard, N., "Bronze Casting and Bronze Alloys in Ancient China," Canberra, 1961.
4. Gettens, R. J., "Proposal for the Preparation, Distribution, and Reporting of Standard Samples for Use in Comparison of Analytical Methods and Sampling Techniques," private communication (1965).
5. Organ, R. M., "The Value of Analysis of Archaeological Objects," *Archaeometry* (1971) **13** (1), 27.
6. Chase, W. T., "Proposal for a Program of Comparative Chemical Analyses on Archaeological Materials," private communication (1970).
7. Youden, W. J., "Statistical Techniques for Collaborative Tests," AOAC, Washington, D.C., 1969.
8. Dozinel, C. M., "Modern Methods of Analysis of Copper and Its Alloys," Amsterdam, 1963.
9. "ASTM Manual for Conducting an Interlaboratory Study of a Test Method," *ASTM Spec. Tech. Publ.* **335** (1963).
10. Ku, H. H., Ed., "Precision Measurement and Calibration," selected NBS papers on statistical concepts and procedures. NBS special publication 300, Volume I (Washington, 1969).
11. Mandel, John, "The Statistical Analysis of Experimental Data," New York, 1964.
12. Flanagan, F. J., "U.S. Geological Survey Standards—II. First Compilation of Data for the New U.S.G.S. Rocks," *Geochem. Cosmochim. Acta* (1969) **33,** 81.
13. Larsen, E. S., "The Accuracy of Chemical Analyses of Amphiboles and Other Silicates," *Amer. J. Sci.* (1938) **35,** 94.
14. Moorey, P. R. S., Schweizer, F., "Copper and Copper Alloys in Ancient Iraq, Syria and Palestine: Some New Analyses," *Archaeometry* (1972) **14** (2), 177.
15. Morrison, G. H., "Evaluation of Lunar Elemental Analyses," *Anal. Chem.* (1971) **43,** (7), 22A.
16. Oddy, W. A., "The Analysis of Gold Coins—A Comparison of Results Obtained by Non-destructive Methods," *Archaeometry* (1972) **14,** (1), 109.
17. Oddy, W. A., Schweizer, F., "A Comparative Analysis of Some Gold Coins," in "Methods of Chemical and Metallurgical Investigation of Ancient Coinage," Hall, E. T., Metcalf, D. M., Eds., Royal Numismatic Society, *Spec. Publ.* **8,** London, 1972.
18. Schweizer, F., Friedman, A. M., "Comparison of Methods of Analysis of Silver and Gold in Silver Coins," *Archaeometry* (1972) **14** (1), 103.
19. Croegaert, J., "Report on Interlaboratory Comparison Experiments on the Analysis of Synthetic Ancient Glasses, Provided by the Corning Museum of Glass," Corning, N. Y., 1966.
20. Dixon, W. J., Massey, F. J., "Introduction to Statistical Analysis," p. 412, Table A-8, New York, 1957.
21. Dixon, W. J., "Processing Data for Outliers," *Biometrics* (1963) **9,** 74.
22. Proschan, F., "Rejection of Outlying Observations," *Amer. J. Phys.* (1953) **21,** 520-525 (*NBS Spec. Publ.* **300** (1969) **1,** 349-354).
23. Langley, R., "Practical Statistics Simply Explained," p. 199 ff., New York, 1970.

RECEIVED July 9, 1973. Adapted from a report presented at the 1972 Madrid meeting of the Conservation Committee of the International Council of Museums, Working Group on Metals.

10

Mössbauer Effect Spectroscopy without Sampling: Application to Art and Archaeology

BERNARD KEISCH

Mellon Institute, Carnegie-Mellon University, 4400 Fifth Ave., Pittsburgh, Pa. 15213

A brief review of Mössbauer effect spectroscopy (MES) is given, especially with regard to iron and its compounds. While most workers have used a transmission geometry which requires sample preparation, MES may also be performed by scattering. Thus, surface analysis is possible with no sampling or damage to an artifact or work of art. A special detector for efficiently accomplishing this has been constructed. We have surveyed various artists' pigments, in bulk as well as in situ *and found that MES leads to a useful classification. Terra cotta statuary, which generally contains 5–10% iron, has also been surveyed. Other workers have examined pottery (using transmission geometry) with interesting results.*

It is well known that iron is a common constituent of works of art and of artifacts of archaeological interest. This is undoubtedly because of its occurrence in most rocks of the earth's crust and in a variety of colored compounds. The chemistry of iron compounds is fairly complex, and the form in which it is eventually found in an object can provide information on the source and the technique of manufacture of that object. Mössbauer effect spectroscopy (MES) is particularly suitable for studying iron and its compounds. Already there are two published reports (*1*, *2*) of MES applied to the study of ancient pottery. However, those studies, like most MES work, required sample taking and preparation. We have explored the possibility of nondestructive MES and now report on practicality of this method. Since there are many good descriptions of MES, at various levels of technical detail, published elsewhere (*3, 4*), only a brief description of the method is presented here.

Mössbauer effect spectroscopy is based on the fact that from certain radioactive nuclides, bound in solids, some gamma radiation is emitted with virtually no energy loss from recoil of the emitting nucleus. The radiation thus emitted has a very sharply defined energy and will interact with other nuclei having the same energy state. The result is an absorption–fluorescence system that is extremely sensitive to any perturbations in the energy states of the nuclei involved. The system sensitivity is so great that it depends, among other things, on the state of the electrons outside of the nucleus and, hence, on the chemical and magnetic state of the material. The resonance interactions can also be affected by inducing a relative motion between the radioactive source nuclei and target (*i.e.*, sample) nuclei. This is a Doppler effect, and for the system, ^{57}Co-^{57}Fe (Figure 1) used to study iron compounds, the velocity of this motion is only of the order of millimeters per second. For a more thorough explanation of the Mössbauer effect, consult Refs. *3* and *4*. Ref. *3* is a discussion of MES for the lay audience while Ref. *4* provides an extensive bibliography in the field.

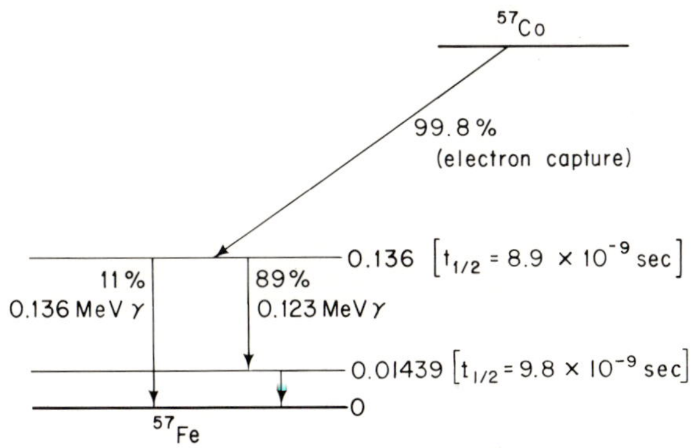

Figure 1. Simplified decay scheme for ^{57}Co–^{57}Fe system

A typical Mössbauer effect spectrometer consists of a transducer which imparts controlled motion of a radioactive source, the sample itself, a suitable radiation detector, and certain necessary electronic instrumentation. Such a system is diagrammed in Figure 2A.

The γ-rays from the source (which must be specially prepared to yield a high fraction of recoilless emission) are absorbed by the sample and re-emitted approximately isotropically. Compared with ordinary electronic absorption of γ-rays by an equivalent material, detector (A) in the figure measures a decrease in transmitted γ-ray intensity when

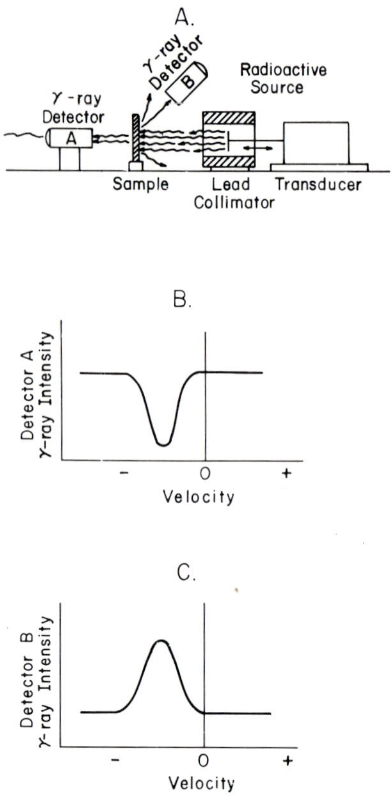

Figure 2. (A) schematic of a Mössbauer effect spectrometer; (B) simple MES spectrum obtained by transmission; (C) simple MES spectrum obtained by scattering

resonance absorption occurs while at the same time detector (B) measures an increase in the intensity of *scattered* γ-rays. Either of these may be used to measure Mössbauer effect. Motion of the source imparted by the transducer can compensate for any effect in the sample which would otherwise yield no resonance absorption or scattering when the source is at rest. This is because the relative energy of the γ-rays is slightly altered by the small Doppler effect thus produced.

Mössbauer spectra are obtained by measuring the transmitted or back-scattered γ-ray intensities as a function of velocity. Figures 2B and 2C show spectra obtained by transmission and by scattering, respectively. The shift of the peak location from zero velocity is typical and is an important parameter often called chemical shift. The amount of chemical shift is a function of the composition of the radioactive source as well.

More complex spectra are usually obtained. Quadrupole splitting results in two peaks equal in intensity to a first approximation. The amount of splitting is an important parameter, with larger splits occurring for Fe(II) compounds than for Fe(III) compounds. Hyperfine magnetic splitting results in a symmetrical six-peak spectrum with a variation in position and in spread related to the chemical state of the compound. Many superimposed combinations of these as well as broadened peaks and other perturbations are observed when complex materials are examined.

The data which yield a spectrum can be obtained with a system in which measurements are made at one velocity at a time (the transducer motion during the measurement is held at constant velocity) or, preferably, by changing the velocity through a range at constant acceleration and electronically sorting the γ-ray intensity data obtained as a function of the changing source velocity.

There is also a temperature effect. Generally at low temperatures (liquid nitrogen or even liquid helium), spectra will show more detail with regard to magnetic properties and are easier to obtain. For the nuclei of many elements, spectra are obtainable only at low temperatures if at all.

Whether or not the effect can be obtained for a particular element depends on a fortuitous combination of half-life and nuclear energy levels. While many elements have yielded such spectra, the system represented by iron-57 (natural abundance approximately 2%) is the easiest to observe, and excellent results are obtained even at room temperature—hence the interest in the method for studying iron compounds in art and archaeology. While most MES data have been collected with transmission geometry, which requires either thin samples or some sample preparation to achieve thinness, data collection by scattering allows one to achieve the same results with no sample preparation whatsoever—*i.e.*, if the compound to be studied lies at or very near the surface of the material in which the compound occurs. For example, in a sample of a typical iron oxide, the analysis would pertain to a surface layer approximately 0.2 mm deep.

In studying works of art and of artifacts the latter proviso is usually satisfied because iron compounds occur as pigments at the aesthetic surface of a painting (often covered by only a thin layer of varnish) or are homogeneously distributed throughout the matrix of statuary or archaeological material.

We have now surveyed the iron-bearing pigments and, to a limited extent, terra cotta statuary using both transmission and scattering geometry. The results provide information of an extraordinary richness and great potential value for identifying and studying the chemistry and physics of the various iron-bearing materials.

Back-Scattering Detector

The ideal back-scatter detector for MES would be one that detects all of the scattered radiation and none of the radiation from the radioactive source—an impossible objective. Other workers (*5, 6*) have determined that, of the three different radiation forms which could be used to obtain a spectrum—*i.e.*, 6-KeV x-rays, 8-KeV γ-rays, and 14-KeV γ-rays—the last is the most efficient for various practical considerations.

The detector of choice for the efficient detection of 14-KeV γ-rays is the gas proportional counter with a (mostly) krypton filling. (Methane, which we use, or carbon dioxide are typically added in a concentration of *ca.* 10% for quenching secondary discharges.)

A window must be provided which can admit this low-energy radiation with a minimum of loss from absorption. Shielding is necessary to protect the detector not only from the bulk of the radiation from the source (which is mainly 6-KeV x-rays, 14-KeV γ-rays, and 123-KeV γ-rays) but also from the small percentage of higher energy γ-radiation ($\sim$ 700 KeV, $\sim$ 0.2% abundance) from ^{57}Co and from the γ-radiation from the inevitable radioisotopic impurities, ^{56}Co and ^{58}Co, with γ-ray energies up to $>$ 3 MeV. All this must be done with a minimum sacrifice of geometrical considerations—*i.e.*, radiation delivered to the sample and back-scattered radiation delivered to the detector.

The detector developed at our laboratories is shown in Figures 3–5. Its design incorporates many features previously explored by others (*5, 6, 7, 8*). An additional important consideration, not mentioned above, is detector resolution. Maximizing resolution means that a greater rejection of background is possible while retaining maximum signal detection. The grid shown in Figure 4 helps achieve good resolution and also eliminates a problem we encountered in which an electrical charge gradually accumulated on the insulators which support the anode.

The window material is beryllium (approximately 0.9-mm thick), the material of choice for low-energy x- and γ-radiation. Aluminum windows can be used but must be strong enough to withstand a vacuum; hence a significant sacrifice in transparency must be made. Suitable plastic windows, restricted to low atomic number materials, leak slightly, and the presence of any oxygen or water will reduce the resolution performance of a detector of this kind. It is essential that the gas fill used be of the highest possible purity and that it remain so.

The shielding used here is primarily tungsten (sp gr = 19.4) which is much more effective than lead (sp gr = 11.3). However, lead is used in the less critical areas for reasons of economy.

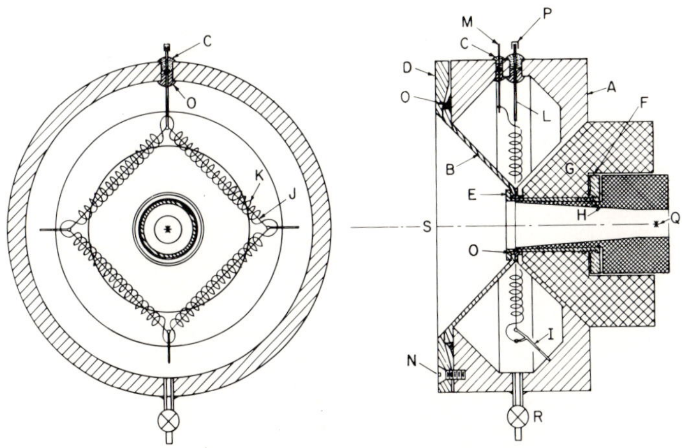

SECTION THROUGH PLANE OF ANODE

SECTION ON A DIAMETER

Figure 3. MES back-scattering detector

A—detector body (aluminum)
B—window (beryllium, 0.034-inches thick)
C—epoxy seal
D—clamping plate (aluminum)
E—clamping sleeve (aluminum)
F—clamping sleeve nut (brass)
G—shielding (tungsten)
H—collimator (tungsten)
I—quartz support (one of three, 2-mm diameter)
J—anode (0.001-inch diameter wire)
K—grid (0.020-inch diameter wire)
L—anode lead (hypodermic needle tubing)
M—grid lead
N—clamping plate bolt (one of eight)
O—high vacuum epoxy seal
P—soldered cap
Q—source position
R—high vacuum valve
S—typical sample position

The result of our effort to develop the best possible detector for MES is as follows. Our detector has a resolution of approximately 2 KeV (fwhm) at 15 KeV as shown in Figure 6. There is virtually no deterioration in performance over a period of several months. The overall efficiency of the detector when used for MES with 14-KeV γ-radiation is such that a 0.001-inch thick sample of stainless steel type 302 (natural isotopic abundance) gives a spectrum with the peak height some 400% of the base line, Figure 7. (For comparison, when we started we were quite content with 50%.) With our 10-mc Co-57 source, the data acquisition rate in the peak is approximately 500 counts/min. This means that in a matter of a minute or less one obtains a recognizable spectrum. As a bonus, the observance of 6-KeV x-rays yields an effect of approximately 50% of the baseline. To accomplish this, we interpose a plastic filter between the source and the sample to absorb most of the 6-KeV radiation from the source (which does not contribute to the effect but is elastically

Figure 4. Two views of the detector without the window in place

Figure 5. The detector complete and in the spectrometer

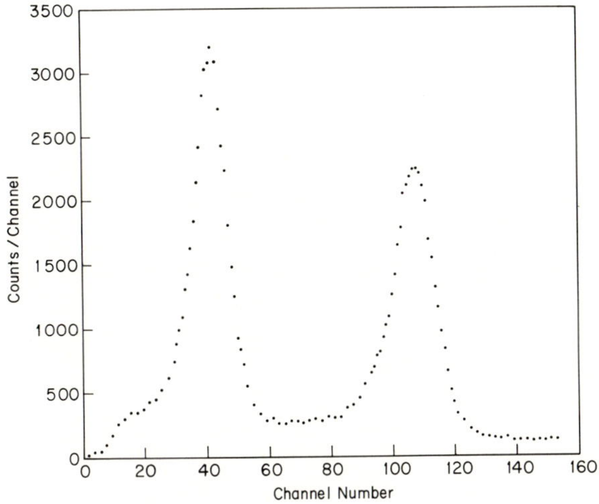

Figure 6. Response of the detector to radiation from ^{57}*Co*

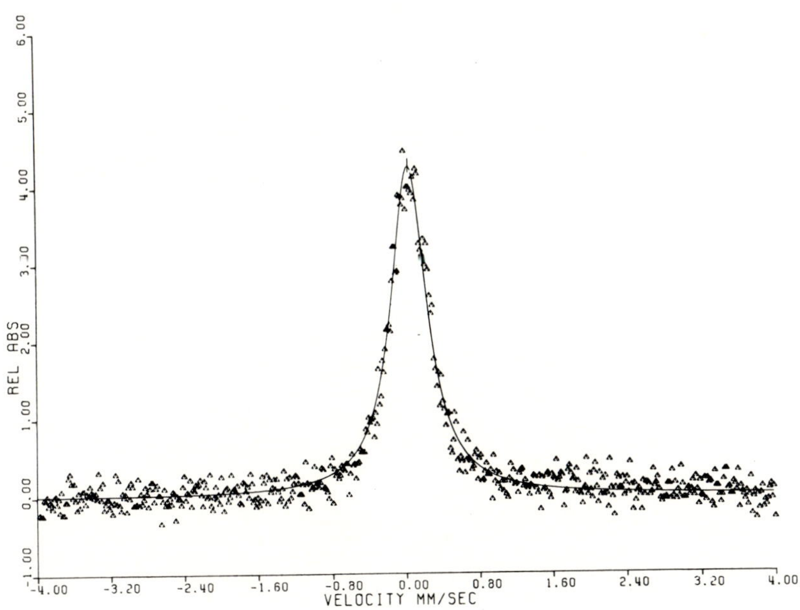

Figure 7. Mössbauer effect back-scatter spectrum for stainless steel (302) with natural abundance of ^{57}*Fe*

scattered). The use of 6-KeV x-rays to observe Mössbauer spectra has an advantage in certain applications when one wishes to analyze a thinner surface layer than usual.

Other Considerations

The electronic equipment used is now commercially available from three or four sources. Ours consisted of a Nuclear Science Instruments Mössbauer spectrometer and a Northern Scientific pulse height analyzer. The spectra we obtained were all fitted (by least squares) by an IBM 360 computer (*7*) used in a time-shared system. Plots were also computer-produced. All of our spectra, whether obtained by transmission or by scattering, are shown with the ordinate as fractional absorption relative to the off-peak background.

Pigments

A fairly comprehensive overview of our survey of iron oxide pigments has been published elsewhere (*9*). Here we briefly indicate the general results to illustrate the nature of MES as an analytical tool.

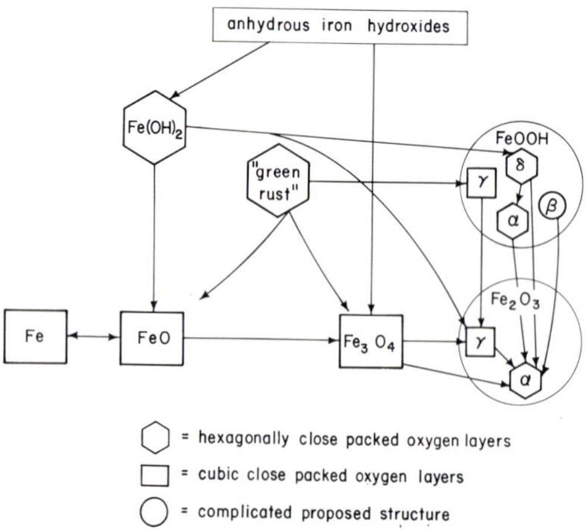

Figure 8. Some iron oxides and their relationships (after Ref. 10)

Figure 8 (*10*) shows the interrelationship of some of the forms of iron oxide. The important pigments are α-Fe_2O_3 (hematite) and α-FeOOH (goethite). While these are generally red and yellow, respectively, there are many variations which we believe arise from particle size distribution and the presence of other minerals. Reduced forms of iron oxide also occur as pigments, notably Fe_3O_4 (magnetite), which are black and the curious pigment known as "green earth" which is a complex mixture of celadonite and glauconite.

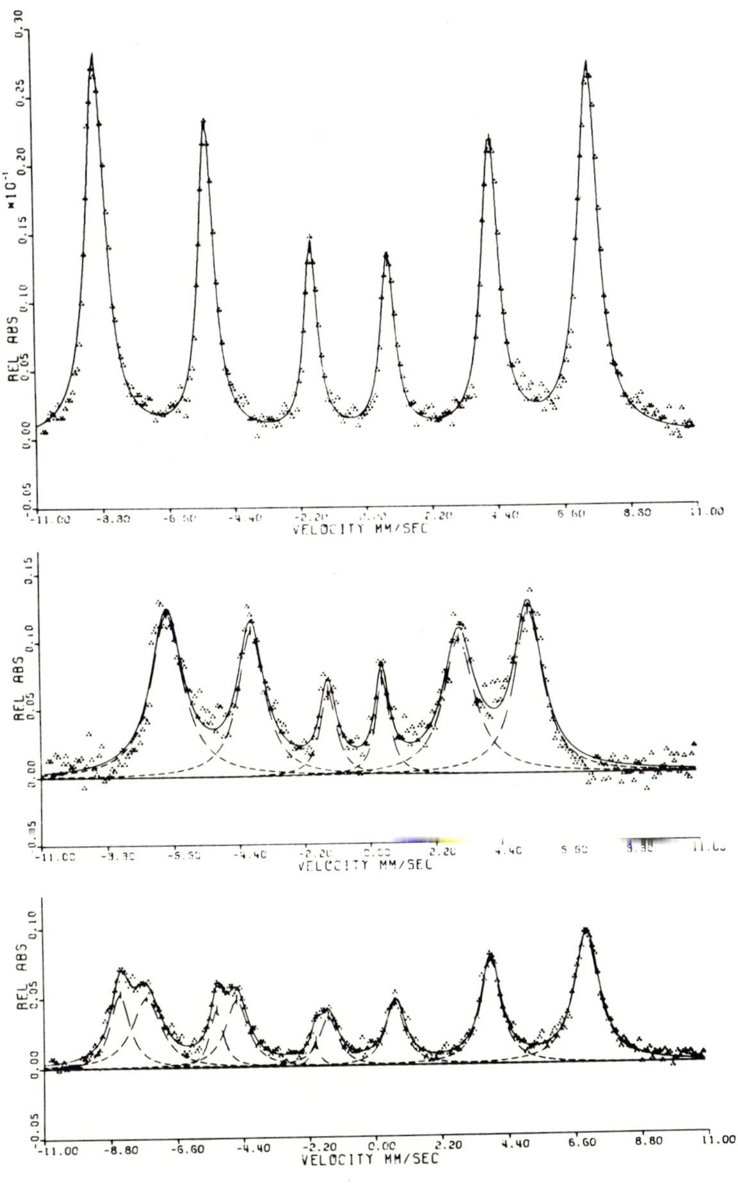

Figure 9. MES spectra for hematite (top), goethite (middle), and magnetite (bottom)

The precise characterization of iron oxide pigments in paintings which, of course, includes ochres, siennas, and umbers, is difficult by ordinary means. This is primarily because of the difficulty of resolving the fine crystals with a light microscope. Characterizing variations within a broad general classification or distinguishing mixtures is virtually impossible. MES, however, offers a wealth of information of this type.

Figure 9 shows spectra for three synthetic materials—hematite, goethite, and magnetite, and obviously there are distinct differences among them. [The ^{57}Co source matrix for the figures shown here was either cobalt oxide (CoO) or chromium. Velocities given are relative motion between the moving source and stationary sample. Exact peak positions are not given here because they are not relevant to this presentation. For additional information on the samples described here and others, consult Ref. 9.] Mixtures, such as a typical synthetic brown (Figure 10), are readily recognized and with some effort at standardization can even be roughly quantitized.

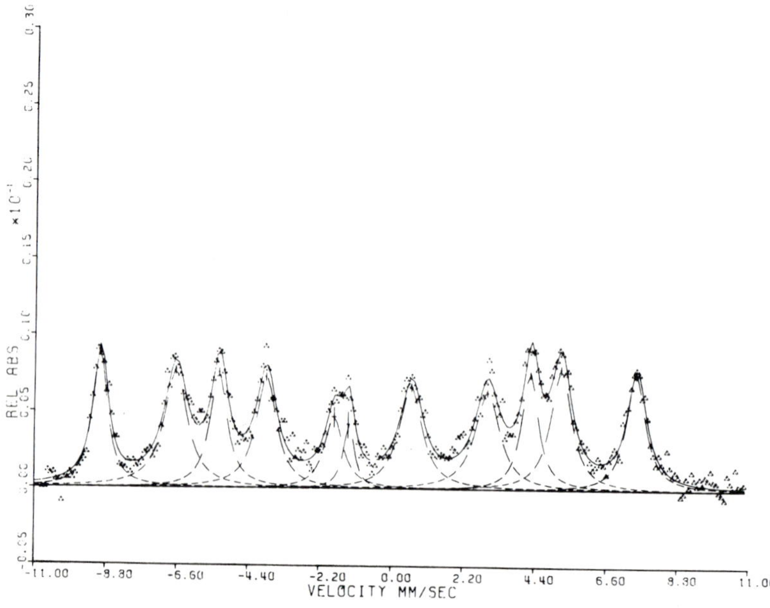

Figure 10. MES spectrum for a synthetic brown (hematite and goethite)

Figure 11 shows the differences between synthetic hematite and one from "burning an ochre" and between a synthetic goethite and a typical yellow ochre (in which the iron is chiefly goethite in character). Note the extra peak or peaks in the center of the hematite spectrum in the natural material.

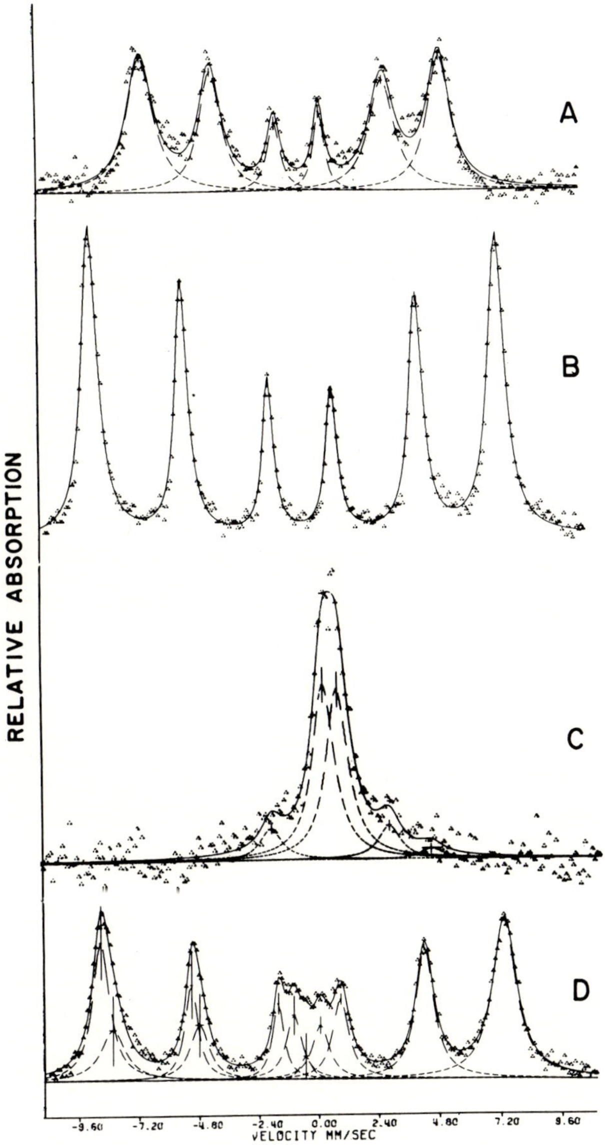

Figure 11. MES spectra for synthetic hematite (A), synthetic goethite (B), yellow ochre (C), and "burnt" ochre (D)

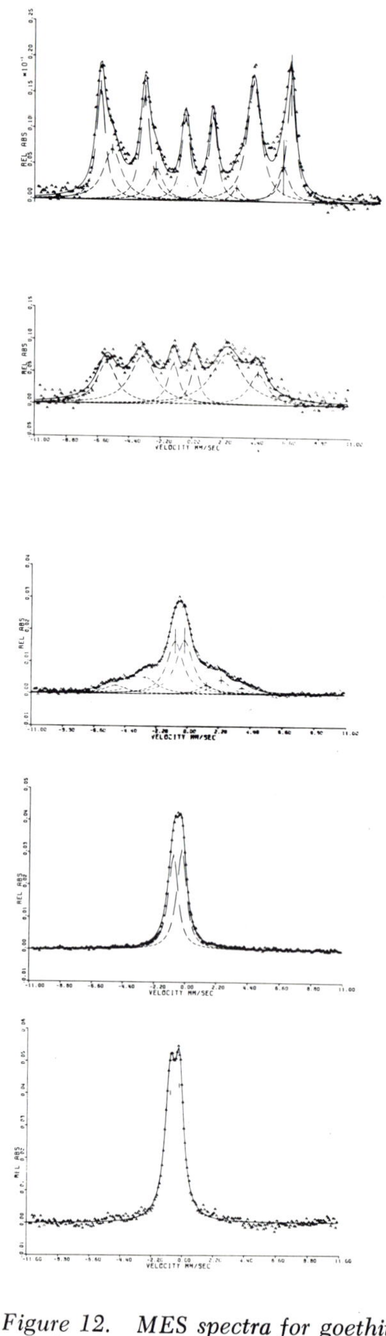

Figure 12. MES spectra for goethite samples showing increasing super-paramagnetism

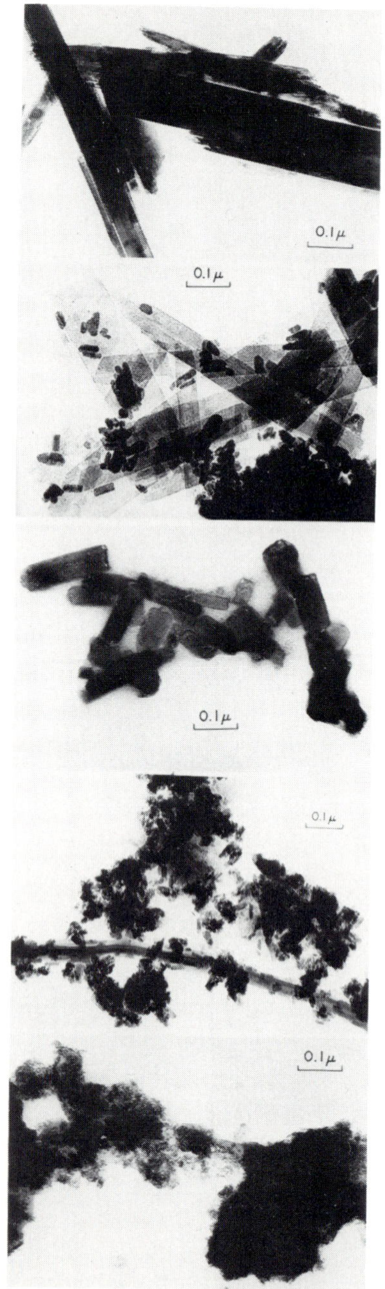

Figure 13. Electron micrographs of the samples in Figure 12

There appear to be two possible explanations for these central peaks in the natural hematite. One is that some of the ferric material is actually not hematite but is bound to clay that might be present. Other workers (*11*) obained similar results when they prepared samples of iron compounds absorbed on kaolinite or bentonite. The second reason concerns the effect of small crystal size on Mössbauer spectra and perhaps is better illustrated for goethite.

Figure 12 shows a series of spectra of pigments which are mainly goethite but of decreasing crystal size. Figure 13 shows electron micrographs of these samples; the correlation is obvious. The progressive collapse of the six-peaked (magnetic hyperfine split) spectra to two peaks has been observed by others (*12, 13*) and explained as "supermagnetism"—a kind of breakdown of the internal hyperfine magnetic field. We have experimentally observed that the "burning" of a finely divided goethite results in a product giving a Mössbauer spectrum like that of a natural hematite while the burning of a goethite with large crystals yields a material with a spectrum like a synthetic hematite. However, in natural materials, clay is also present in almost every case. The study of mineralogical iron-bearing materials is extremely complex, and often no specific chemical–crystallographic structure is identified. Nevertheless the intercomparison of data is quite valuable for classification because general classes are identifiable.

In our survey we observed that, in general, natural ochres are more finely divided than synthetic ones and that this property also varies somewhat with the source of the material. Of course, the Mössbauer spectrum shows this quite clearly. We also observed other variations in the spectra of such materials as umbers and siennas. Two typical spectra are shown in Figure 14.

This is the kind of information that one may obtain, using the detector without sampling of a painting. Figure 15 shows the setup for analyzing a painting in which a tan ribbon is aligned with the center of the detector. The spectrum obtained is shown at the bottom and is that of a typical natural yellow ochre.

Figure 16 shows the spectra obtained for two other pigments—green earth and Prussian blue. In the former case there are actually two spectra, one of which consists of a widely split doublet typical of iron(II) compounds. Heating this pigment changes it to a brownish-red and removes the ferrous doublet.

The spectrum for Prussian blue also consists of two spectra. One is from the $Fe(CN)_6^{4-}$ ions, and the other is from the Fe^{3+} ions. We are currently exploring this system using MES to investigate the process of fading. The objective would be two-fold—first to be able to identify objectively faded Prussian blue in a painting, and second to study the possibility of reversing the fading.

Terra Cotta

Terra cotta typically contains approximately 5–10% iron. We surveyed approximately 20 specimens of sixteenth-to-nineteenth century terra cottas and observed two general types of Mössbauer spectra (Figure 17). In some samples we observed a combination of these, and this led us to heat a sample to progressively higher temperatures and analyze it. Figure 18 shows the results. Note the transition between 470°C (actually a similar spectra was obtained after heating at ~ 800°C) and 1000°C. At 1200°C, for which the top curve was observed, the material had melted to a blackish-brown glassy material. While the number of samples we examined is still too limited to draw firm conclusions, we

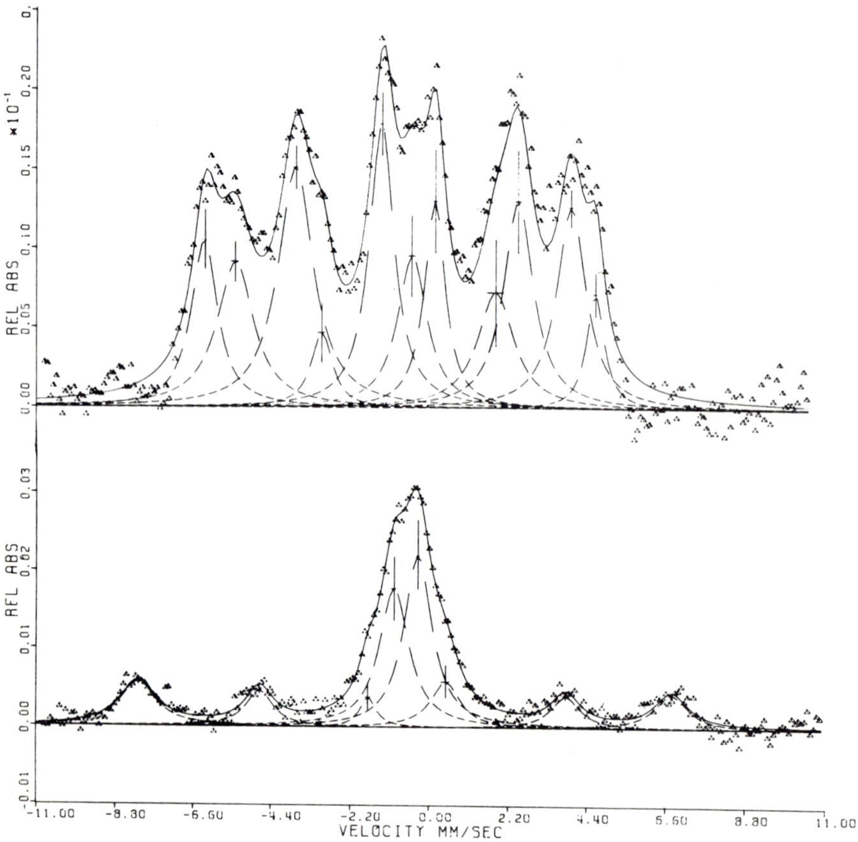

Figure 14. MES spectra for a sienna (top), and an umber (bottom)

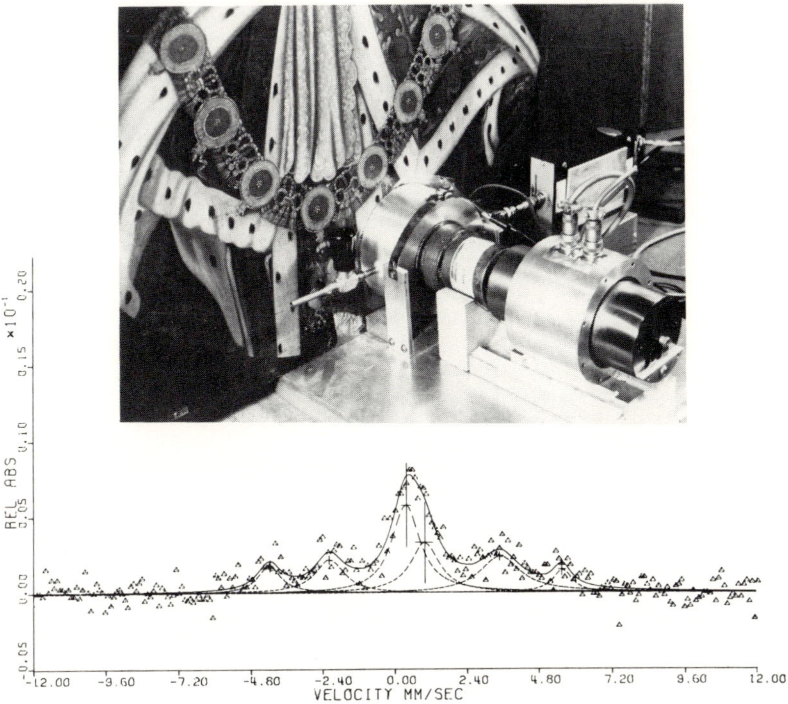

Figure 15. MES of an area on a painting

may tentatively state: (a) earlier work (early Renaissance) appears not to have been heated to quite as high a temperature as later work, (b) sometimes one may distinguish one particular artist's work, and (c) forgeries that have been prepared by coloring a mortar or cement-like material can be readily distinguished. Spectra from such mixtures indicate that a pigment was used to obtain the color of genuine terra cotta.

A few practical applications have already resulted from the application of this method. In one work, where an arm had been pieced onto a statue, MES clearly showed that the arm was of a different material, and thus no sampling was necessary. In a coordinated group of three objects, it was clearly shown that one of the objects, presumably recent, did not belong in the group. Also, among our samples were two objects of definite attribution to one artist and a third work of questionable attribution. MES clearly showed that the third work was of a different composition.

Pottery

Two groups of workers have used MES to study ancient pottery. In both cases, samples were taken, and transmission spectra were obtained. In one study by Cousins and Dharmawardena, the samples examined were taken from sherds and clays found at or near Cheam in Surrey (England) (*1*). A buff sherd (P18) with light sand tempering gave the spectrum shown in the upper part of Figure 19. This spectrum is similar to some of those we obtained from terra cotta. A sample of clay from that area gave similar spectra when baked but with a regular variation in spectrum parameters as a function of baking temperature.

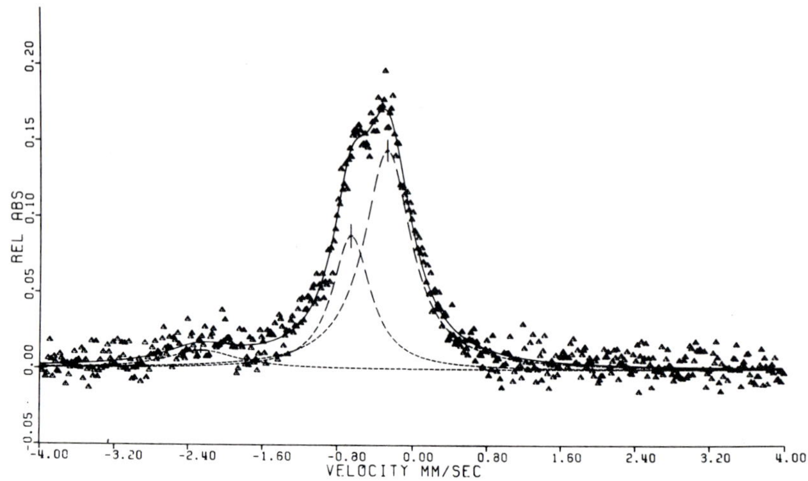

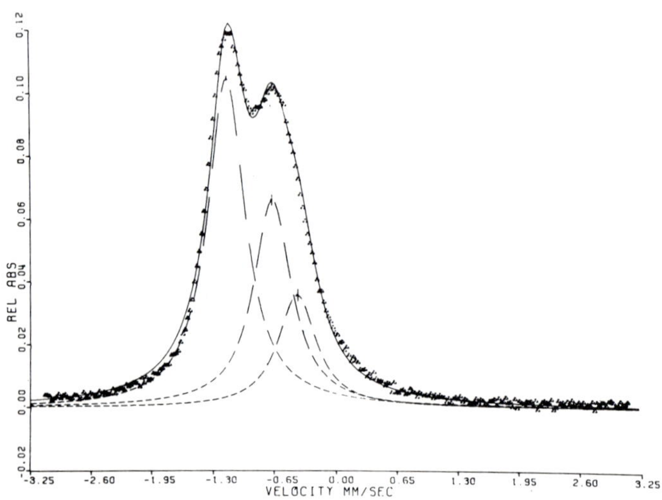

Figure 16. MES spectra for green earth (top), and Prussian blue (bottom)

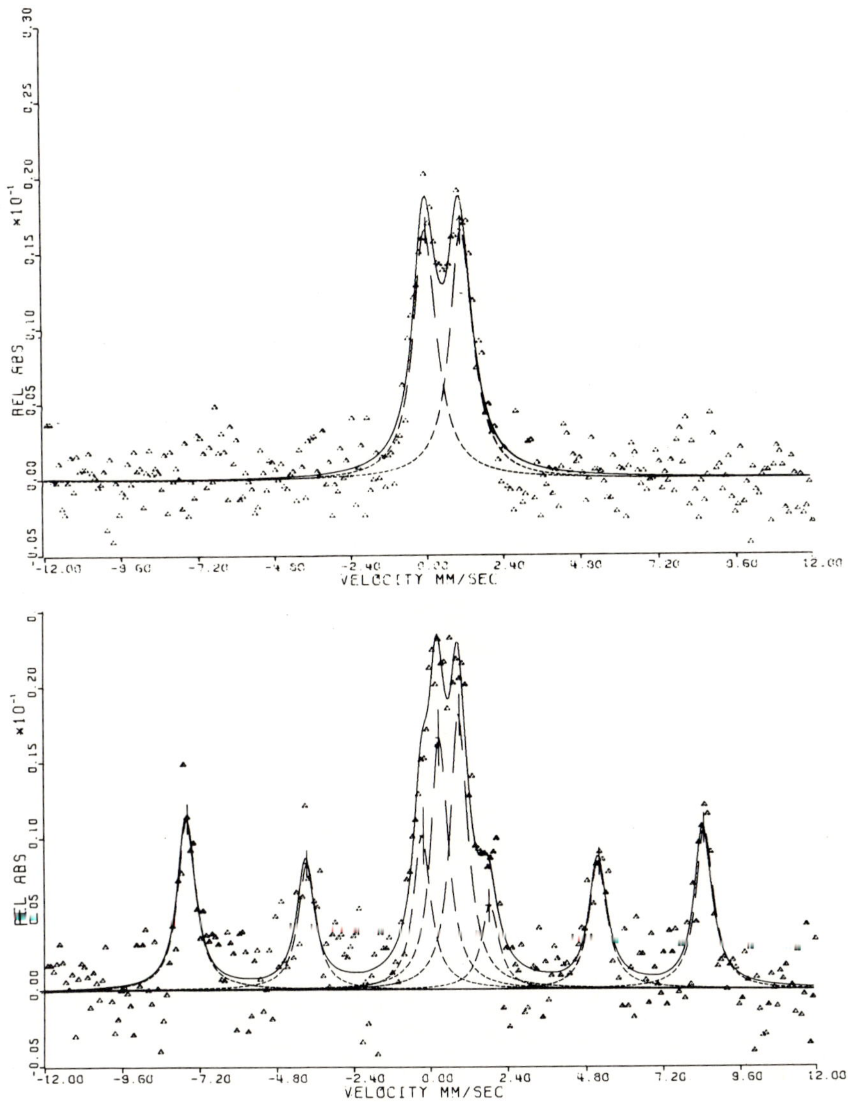

Figure 17. MES spectra for two "types" of terra cotta

The authors indicate that this can lead to an estimate of baking temperature. They also point out that superficially similar pottery found about 10 miles away (Kingston) yielded entirely different spectra.

Cousins and Dharmawardena also examined a lightly gritted black ware and obtained the spectrum shown in the lower part of Figure 19. This shows the presence of both Fe(II) and Fe(III). Baking this sample in air at 700°C yielded a buff material that was entirely Fe(III) "but not oxide." (The new spectrum was not quite like that of P18). A red sandy ware with a black interior was also examined. The differences between the red and black materials were clearly shown to be differences in oxidation states. Only 27% of the iron in the black form was in the Fe(III) state.

Gangas *et al.* (*2*) made a similar exploratory study of ancient Greek pottery. They prepared thin samples by scraping 200 mg from the insides of the pottery. They found that there were distinct differences in the spectra obtained which were correlatable with archaeological evidence. Among four main groups of pottery examined, clear mineralogical differences were obtained for samples that were the same in appearance.

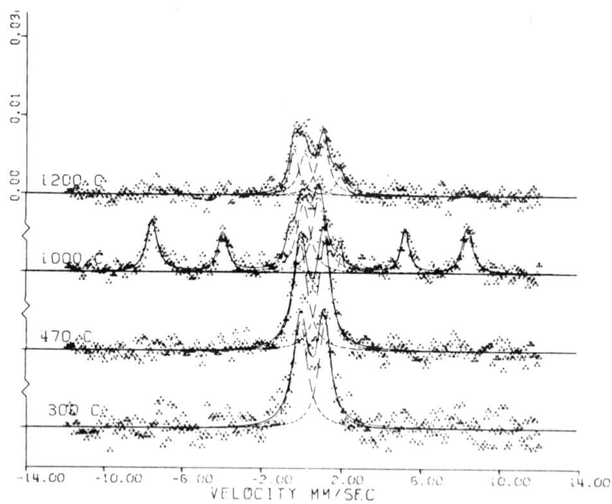

Figure 18. MES spectra for a sample of terra cotta heated to various temperatures prior to analysis

The Mössbauer spectra obtained even divided one group of Korinthian pottery into two subgroups which could be correlated with date and style.

Summary

Back-scatter Mössbauer effect spectrometry can be an effective non-destructive tool for characterizing iron compounds that occur so frequently in works of art and in archaeological materials. In paintings, this technique can be used to identify various iron-bearing compounds to aid the restorer–conservator in discerning which parts of a work are original and how to match them. For identification, MES can yield valuable clues as to the origin of a work by discerning between synthetic

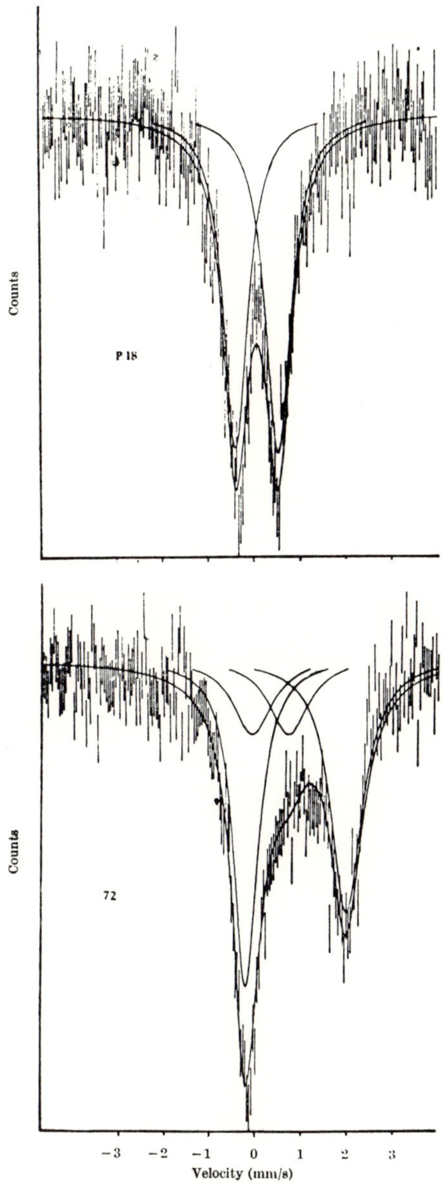

Figure 19. MES spectra of two potsherds (after Ref. 1)

and natural materials. Terra cotta can be classified to help with distinguishing artists, periods, and sources. Finally, pottery, both whole and sherds, can be classified as to period, location, and perhaps firing temperatures. While information obtained by this method may not always be conclusive or unambiguous, nothing is lost by making the analysis because of the absolutely non-destructive nature of this technique. In effect, it is less destructive than ordinary x-radiography that is so commonly used.

Acknowledgments

The work presented here would not have been possible without the help of: The United States Atomic Energy Commission, Division of Applied Technology, Contract AT(11-1)-3034, the National Gallery of Art (Washington, D.C.), P. A. Flinn of the Carnegie-Mellon University, Physics Department, Robert L. Feller, and the Pigment Bank of the Art Research Project at Carnegie-Mellon University, and various pigment manufacturers.

Literature Cited

1. Cousins, D. R., Dharmawardena, K. G., *Nature* (1969) **223**, 732.
2. Ganges, N. H. J., Kostikas, A., Simopoulos, A., Vocotopoulou, J., *Nature* (1971) **229**, 485.
3. Goldanskii, V. I., *Int. Sci. Technol.* (Dec. 1968) 40-48.
4. "The Mössbauer Data Index," Stevens, J. G., Stevens, V. E., eds., IFI/Plenum, published annually.
5. Koch, R. C., Chow, H. K., Bogner, R. L., "Mössbauer Spectrometry for Analysis of Iron Compounds," *U.S. At. Energy Comm., Rept.* **NSEC-4010-1** (1967).
6. Chow, H. K., Weise, R. F., Flinn, P. A., *U.S. At. Energy Comm., Rept.* **NSEC-4023-1** (1969).
7. Flinn, P. A., O'Connell, T., *U.S. At. Energy Comm., Rept.* **WASH-1220** (1973) 92.
8. Keisch, B., *Nucl. Instr. Methods* (1972) **104**, 237.
9. Keisch, B., *Archaeometry* (1973) **15**, 1, 79.
10. Bernal, J. D., Dasgupta, D. R., Mackay, A. L., *Clay Minerals Bull.* (1959) **4**, 15.
11. Yassaglou, N. J., Peterson, J. B., *Soil Sci. Soc. Amer. Proc.* (1969) **33**, 967.
12. Shinzo, T., *J. Phys. Soc. Japan* (1966) **21**, 917.
13. Van der Krann, A. M., Van Loef, J. J., *Phys. Lett.* (1966) **20**, 614.

RECEIVED July 9, 1973.

11

Preliminary Investigations on Chinese Ink in Far Eastern Paintings

JOHN WINTER

Freer Gallery of Art, Smithsonian Institution, Washington, D.C. 20560

Chinese ink, an intimate mixture of combustion carbon and animal glue, was originally based on pine wood soot, but lampblack from oil lamps became dominant after about the 11th century. Various inks, many from paintings, were studied by scanning electron micrographs; these enable individual carbon particles to be seen and some approximate particle size distributions to be constructed. Combustion carbon can be recognized easily as a pigment in paintings. Lampblack inks seem to have a fairly narrow particle size distribution; pine soot inks give more variable distributions. The reason some inks show a bluish and some a brownish tint is concluded to be light absorption differences by the carbon rather than additives or Tyndall effects. Though falling into the same range, distributions differ sufficiently in shape to suggest that they could be an identifying characteristic in some cases.

Writing and the written word have always had a special importance in Chinese culture. Paper and printing are both Chinese inventions; the hair brush, also used for writing, has a considerable antiquity in that country. Although black inks and paints were made in other parts of the world, perhaps from an earlier time (*1*), it was undoubtedly the importance of literacy among the Chinese that led them to produce ink of a quality unmatched anywhere outside the Far East. Furthermore, unlike most other parts of the world, this same material was used not only for writing and printing, but it became almost ubiquitous in painting. In the present context there is no distinction, of the type found in western practice, between ink and paint. Chinese ink could well be described as black water-color paint; it was applied with a brush and used the same glue vehicle as that used with other pigments. The experimental results offered here are derived from samples taken from paintings rather than from examples of printing or writing. However, Chinese ink is a term commonly understood in several different fields, and it would be confusing and pedantic to replace it.

Few Chinese, Japanese, or Korean paintings on flexible supports do not use black ink to a considerable extent. Many were executed using only this medium or with slight addition of color. The virtual ubiquity of ink is undoubtedly one factor behind its importance in the Far East; this is shown by the respect accorded its makers, by its treatment in Chinese historical writing, and by the extensive ink lore that exists.

This paper describes the preliminary and exploratory phase of an investigation of Chinese ink in Far Eastern paintings—its properties and its variability. To clarify what such investigation might achieve, we must first look briefly at the history and method of manufacture of the ink. Note that Chinese ink has been variously called China ink, India (or Indian) ink, Japanese ink, and occasionally by its Japanese name, *sumi.* The Chinese word, less used, is *mo.*

History and Manufacture

Chinese ink is essentially a mixture of soot or lampblack with glue; other constituents may be involved and are mentioned below. The established manufacturing method stresses both the careful choice and preparation of the ingredients and the assidousness and diligence required during their compounding.

The origins of Chinese ink are obscure. There are references to the early use of a black varnish (*2*), an idea that has been disputed (*3*). An early material called *shih mo,* or "stone ink" also crops up in Chinese literature (*2, 4*) and may have been graphite or coal. Archaeological material seems to be sparse and scarcely studied, although an investigation (*5, 6*) on the black pigment found on a Shang oracle bone suggests that it is carbonaceous and possibly blood that has darkened with time. Some references exist to pre-Han potsherds bearing black ink apparently applied with a brush as a liquid (*4, 7*).

Ink as we now recognize it is traditionally believed to date from the Wei dynasty, and the first extant description comes from the 5th century A.D. (*8*). [Ref. 9 erroneously gives the Wei period as "220 à 260 avant J.C." where "après J.C." is clearly meant. Accepted dates for the Wei are A.D. 220–265.] A considerable period of development before then seems probable, however. Since that time the Chinese literature on ink has been fairly extensive; a detailed handbook on inkmaking in the 14th century has been translated into French (*9*), and more recently Franke has rendered many texts into German (*4*). These and other authors (*2, 10–15*) have described the history of inkmaking as derived from the Chinese literature.

From the 5th century the method for making ink remained unchanged in essentials. The carbon was made in ways which did vary, and this aspect is dealt with in more detail below. Animal glue was prepared using collagen from various sources, favorites being deer antler and skins of various animals although fish glue is also mentioned. All accounts stress the assiduousness with which the soot or lampblack, finely sieved before use, and the solution of glue should be mixed. Kneading, pounding, steaming, and rolling are all recommended, usually in combination, and are to be done repetitively and at length to obtain a good product. It may be inferred that properly made ink will have finely divided carbon (a hydrophobic material) and protein glue (a hydrophilic material) intimately mixed—a point which is of some interest regarding its behavior.

Besides carbon and glue various other constituents have been quoted as being incorporated by this or that inkmaker, some of them for obscure reasons. Franke (*4*) states that up to 1100 are known; they include perfumes like camphor and musk, dyes (evidently to improve the appearance in some way), plant extracts, and other materials, some of which may have improved the colloidal behavior of the system.

What concerns us is the carbon pigment itself. This has been prepared in different ways, and the differences have historical significance (*2, 4*). Through T'ang times or later the standard preparation was the controlled burning of pine wood in long ovens of bamboo covered with paper. Woodcut illustrations show at least two oven designs that were used (*4, 13*); some are as long as 100 feet. The best soot for inkmaking was collected at the end remote from the fire; carbon from too close to the fire was considered unsuitable. This collection method would mean that the soot proper was separated from unwanted debris such as fly-ash, charred fragments, and tarry products. Soot particles may also have been fractionated, smaller ones tending to get into the better inks. Woods other than pine were evidently tried but found unsatisfactory, and pine itself was graded for inkmaking suitability (*4*). In any event, this method of obtaining the carbon was probably inefficient.

At some time, perhaps around the 11th to 12th centuries, the production of carbon by burning oil at a wick became important. According to Laufer (*2*), and especially to Franke (*4*), oil burning was associated with the extensive deforestation of China during the first millenium A.D.; it resulted in a shortage of pine wood as raw material, and it coincided with an increasing ink demand for clerical and similar purposes. In any event lampblack production from oil is probably more efficient than pine soot production (though admittedly using a more expensive fuel), is easier to control, less vulnerable to the elements and to fire hazards, and requires less elaborate equipment. Conveniently, if arbitrarily, we may refer to carbon from wood burning as soot and that from oil burnt in a lamp as lampblack. The relative quantities of the two types produced when both were in use are difficult to judge, and the literature is conflicting on this point (*4, 13*). Other fuels for producing carbon are mentioned from time to time, among them petroleum (*4*) and pig's fat (*11, 12*). Their importance is hard to assess, but most authors regard it as minor.

The ink stick also evolved into a minor art form. Many sticks exhibit surface decoration based on varied motifs (*4*); later they were molded into sculptured forms (*3*). The latter were presumably made especially for collectors rather than for use.

Chinese ink was made in Korea from an early date; at the beginning of the T'ang period that country remitted an annual tribute of pine soot ink to the Chinese court (*2*). The technique of making ink seemed to be introduced to Japan by an envoy sent by the King of Korea in 610 A.D. (*2*).

A few technical studies on Chinese ink have been published before. One of the earliest (*16*) included a brief investigation of electrophoretic properties which showed that ink particles exhibited an isoelectric point between pH 4 and 5. This corresponds well with the isoelectric point of gelatin [*ca.* 4.5 (*17*) and *ca.* 4.8 (*18*)] and shows that the carbon particles have an adsorbed layer of collagenous protein from the glue that is known historically to be present. Gelatin or glue is an efficient colloid protecting agent (*19*), and the properties of ink suspensions as colloidal systems are expected to be those involving particles with a protein surface rather than a carbon surface.

Two other papers have studied ink from the point of view of its colloidal properties: one with special reference to the technique of *suminagashi* (*20*) and one with reference to its dispersion by rubbing on inkstones (*21*). A Chinese paper gives some general data (*22*). Two papers (*23, 24*) show electron micrographs of ink particles with, in one case (*23*), observation on the tint and the origin of the carbon (*see* below). A book primarily on inkstones also treats ink in some descriptive detail (*25*).

Experimental

The results presented here are based on scanning electron micrographs (SEM) of paper fibers coated with ink. The instrument used was the Stereoscan Mark IIA (Cambridge Instrument Co., Ltd.), owned by the National Museum of Natural History. A sample consisted of a single fiber excised from an ink-coated area of paper (usually a painting) and mounted on a 12-mm circular glass cover slip by securing one end with poly(vinyl acetate) emulsion. The slip was then stuck to an aluminum stub and coated with 20–30 nm of gold before scanning.

The paintings sampled are listed in Table I. In addition, two modern ink sticks were used. One (designated modern lampblack ink) was a stick kindly given to the Freer Gallery by the Kobaien Ink Works in Kyoto and known to have been based on lampblack from vegetable oil lamps. The other (modern bluish ink) was purchased commercially (Aiko's Art Materials Import, Chicago) and, according to a leaflet supplied with it, was made from pine soot. These two inks were rubbed up with water, daubed on to *hōsho* (an absorbent Japanese paper), and the daubs were sampled as for a painting. The remaining sample was of household chimney soot, generated by a wood-fueled fire, which was mulled with gelatin solution and spotted on to the cover slip.

Table I. Paintings in the Freer Collection Sampled for Scanning Electron Microscopy

Figure No.	*Artist*	*Country, Period*	*Subject*	*Accession No.*
6 7 8	Tsou Fu-lei	China, 14th cent.	plum branches in flower (makimono)	31.1
9	Wang Hui	China, 17th–18th cent.	landscape (makimono)	50.19
10	Kung Hsien	China, 17th cent.	winter landscape	61.11
11	Wang Yuan-ch'i	China, dated 1706	river landscape	62.5
12	Fujiwara Nobuzane (attrib.)	Japan, 13th-14th cent.	portrait Minamoto no Kintada	50.25
13	Chŏng Sŏn (attrib.)	Korea, 18th cent.	landscape (album leaf)	60.9

The particle size distributions (Figures 1–5) were generated as follows. A contact print of a 4 $\times$ 5 inch SEM negative at a magnification of 10^4 was covered with a transparent plastic sheet, ruled with a grid of 1-cm squares. Each square was examined using a 6$\times$ magnifier having a reticle engraved with a 1-cm^2 particle measuring grid. The diameter of each identifiable single carbon particle was estimated to the nearest 0.01 cm and read into a tape recorder. Later the tape was played back and transcribed. At $10^4\times$, the diameters on the micrograph in centimeters are equivalent to actual diameters in micrometers. For plotting, the diameters were grouped in pairs (*i.e.*, 0.05 + 0.06 μm, 0.07 + 0.08 μm, etc.), and the numbers were normalized to a total of 100 particles. Those of diameter less than 0.05 μm were grouped and plotted as a combined group.

In all of the plots except one (Figure 1), analyses of at least two micrographs on samples from the same painting or ink are combined. The lower (top of the shaded area) and upper (top of the open area)

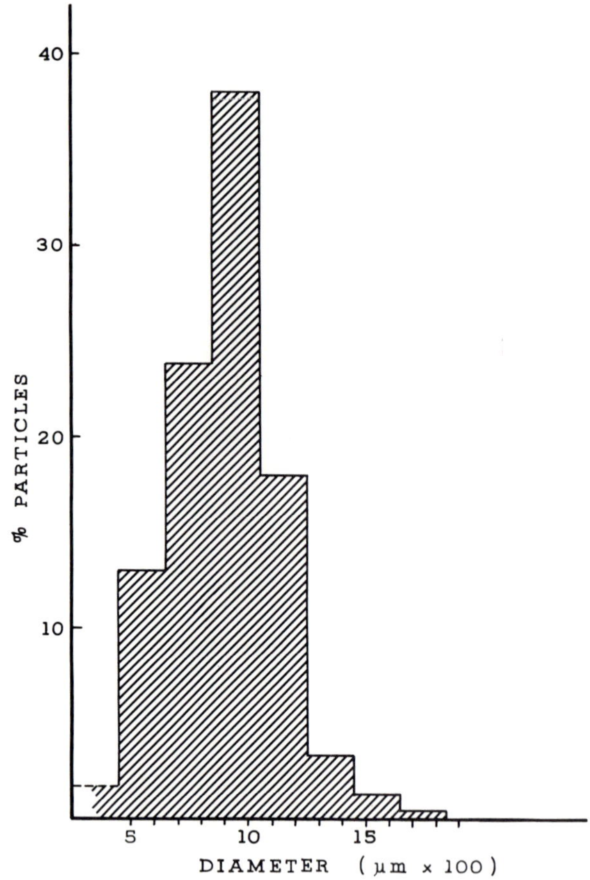

Figure 1. Particle size distribution; modern lampblack ink. Plotted from one analysis.

limits drawn for each class of particle diameters represent respectively the least and greatest numbers of particles in that class (as percentages) that were found. Thus the plots give some idea of the variation over different determinations as well as of the overall shape of the distribution.

Experiments on the modern bluish-tinged ink in connection with the discussion on ink tints were as follows:

(a) Powdered ink (0.11 gram) was Soxhlet extracted successively with acetone (4 hrs), 2-propanol (2 hrs), and pyridine ($6\frac{1}{2}$ hrs). No colored material was extracted.

(b) Ignition of ink in air to constant weight: ash residue, 1.5%. Semiquantitative analysis of ash by emission spectrography gave the following elements in concentration $> 1\%$: Al, Ca, Fe, Si. Ignition of a typical modern brownish-tinged ink gave an ash (0.8%), similar analysis of which showed the same four elements in concentration $> 1\%$.

Results

Precision and Accuracy. This preliminary stage was done to estimate the ways in which particle size distributions vary and to see how far they can be utilized in practical ways. The above technique was chosen to enable us to obtain some approximate results conveniently, without elaborate sample preparation. The scanning electron micrographs (Figures 6–15) show that too great an accuracy should not be expected at this stage, especially in view of the ill-dispersed nature of the particles in some cases (Figures 7 and 10).

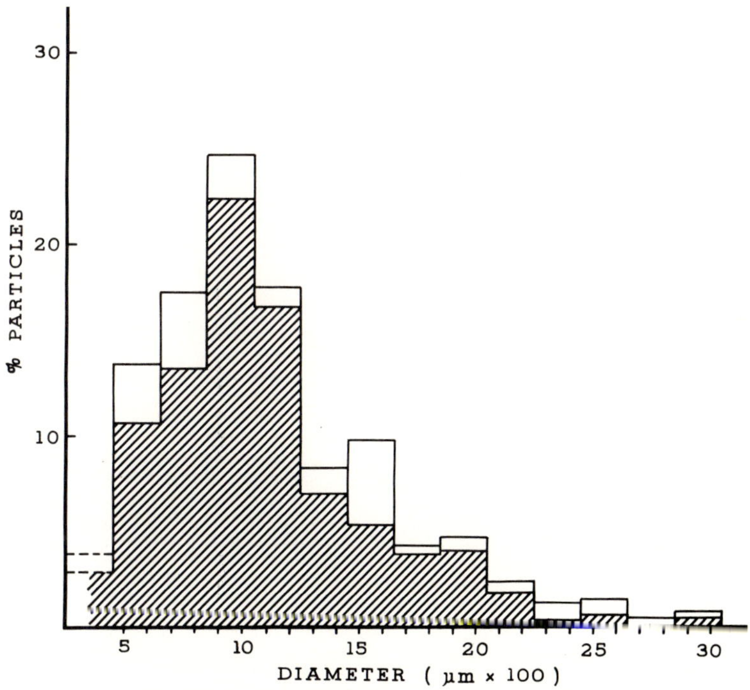

Figure 2. Particle size distribution; modern bluish-tinged ink from pine soot. Plotted from two analyses.

The instrument's resolving power is specified to be 0.025 μm or better. This limitation alone dictated the grouping of particles in ranges of 0.02 μm before plotting. Reproducibility of the instrument's magnification factor was within ±5%. A slight systematic bias towards larger diameters may be present because of the gold coating on the samples, but it should be about constant over this series.

The greatest sources of error come from the difficulties in estimating diameters accurately in these micrographs. They include: misjudgement

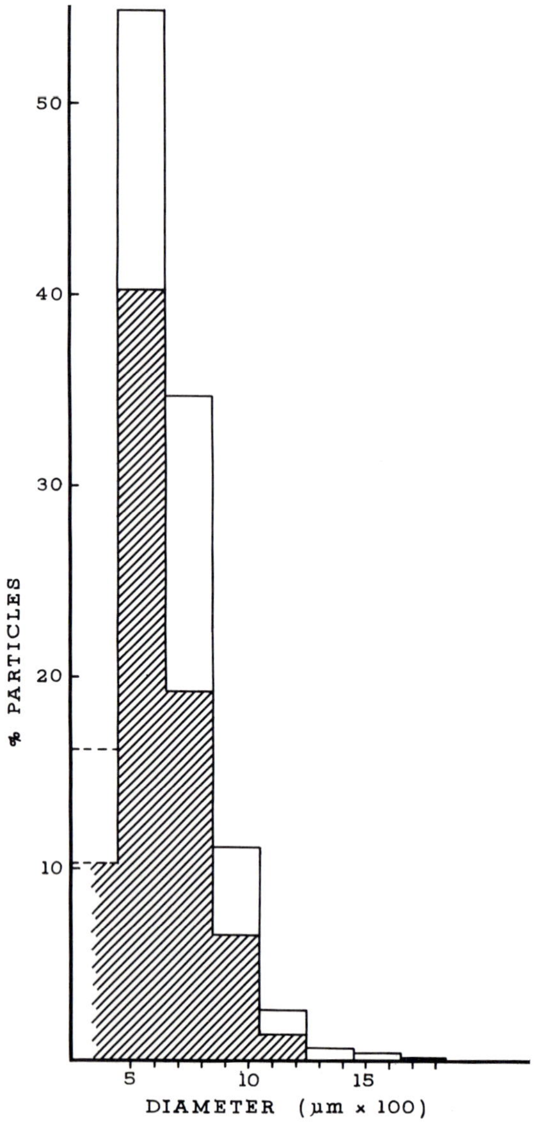

Figure 3. Particle size distribution; 14th century Chinese painting No. 31.1. Plotted from five analyses.

of a clump of particles as one single one; underestimation of the diameter because of partial embedding of a particle in glue or in a clump; selective hiding of particles by others (smaller particles may be hidden more easily than large ones). In addition, it is difficult to avoid a subjective

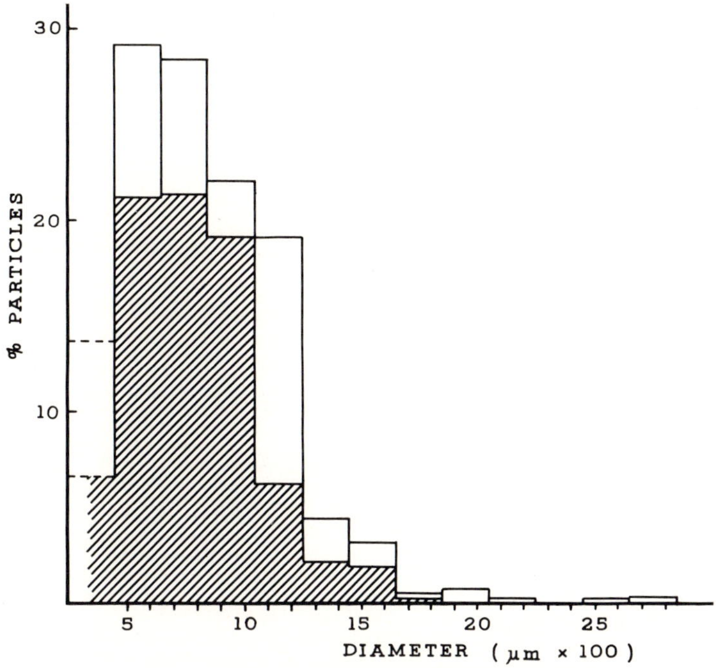

Figure 4. Particle size distribution; 14th century Japanese painting No. 50.25. Plotted from three analyses.

selection factor; the operator must continually decide whether or not a particle is measurable and must do so without biasing the selection for or against a particular size.

The reproducibility of the distribution was checked statistically in several cases—*e.g.*, the Chinese painting 31.1 (Figures 3, 6, 7, and 8). Five micrographs from three samples (three were from one sample) were analyzed. All gave the same kind of distribution—a strong peak in the 0.05 + 0.06 μm class, falling rapidly to zero with increasing diameter (Figure 3). This is visually confirmed by the micrographs (Figures 6–8) which show many small carbon particles and virtually no larger ones. Comparison of the plots in pairs by a conventional χ^2 test showed variable results, ranging from no significant difference to significance at the 0.1% level. Where the difference was significant, it was caused mainly by shifts of particles between adjacent size classes—a natural result of inaccuracy in the diameter measurement. Similar checks on other samples confirmed this impression. Figures 6 and 7 show different degrees of particle matting, leading to possible bias as noted above. Results from these two micrographs, which were from the same sample, were significantly different. Although the gross features of these particle size distributions may be trusted, the present ones are too inaccurate for close statistical analysis.

Discussion

General Features of Chinese Ink. Figures 6 to 13 show a representative selection of scanning electron micrographs of inks found in oriental paintings. The carbon particles are spherical and have diameters up to about 0.20 μm.

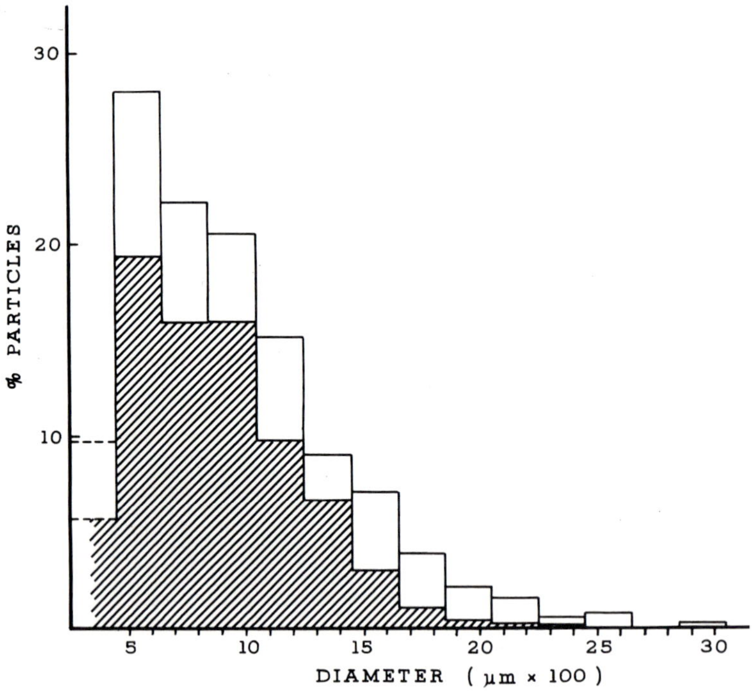

Figure 5. Particle size distribution; 18th century Korean painting No. 60.9. Plotted from four analyses.

According to the literature (*26, 27*), soot from diffusion flames contains spherical carbon particles from 0.01 to about 0.20 μm in diameter. The structure of such a flame shows a number of zones; in the inner zones carbon is formed from gaseous fuel molecules; the outer ones see the subsequent combustion of the particles. Where the latter process is incomplete, the flame produces soot. We will call such material combustion carbon to distinguish it from that derived by the pyrolysis of existing solid, involatile fragments, which we call char. A char consists of irregular particles of various sizes, some of which may

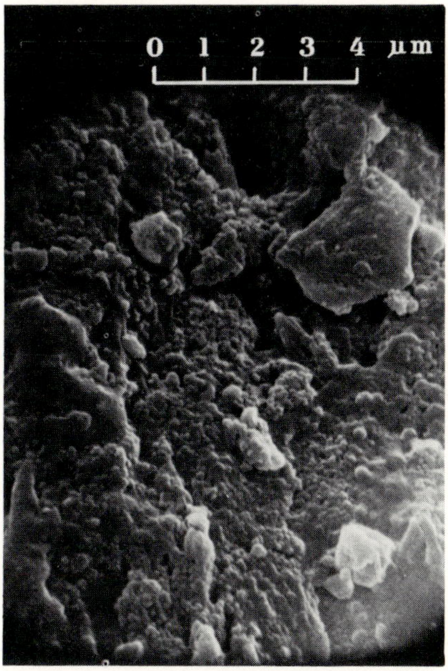

Figure 6. Scanning electron micrograph; ink on paper fiber from 14th century Chinese painting No. 31.1

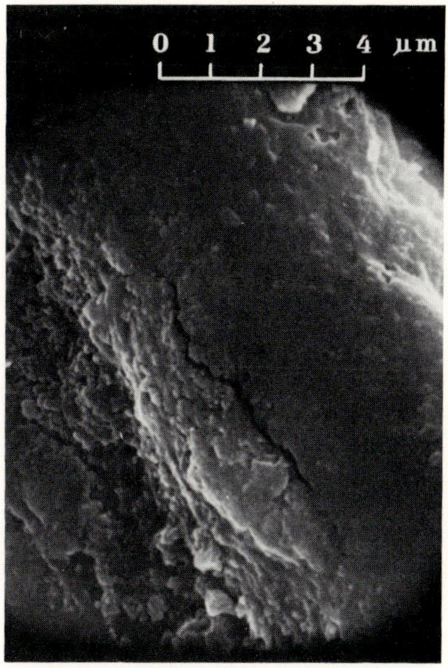

Figure 7. Scanning electron micrograph; ink on paper fiber from 14th century Chinese painting No. 31.1

reflect the morphology of the burnt material. Figure 16 shows a particle of household chimney soot which is itself a charred fragment having combustion particles attached to, and partly fused into, the surface. Although chars are possible in soot inks, the design of soot ovens ensures considerable fractionation of the carbon, only that which settles least rapidly being used. This discriminates heavily against chars, and no such material has yet been positively identified in paintings.

The particles in our samples are certainly of combustion carbon. They may be well dispersed (Figure 11), partially clumped or matted by the glue medium (Figure 9), or heavily matted by the medium (Figure 10). No association of this last feature with schools or periods of painting has been detected, except that paintings on silk usually have more heavily aggregated ink than those on paper. Indeed, the preceding three examples were all from closely related paintings (17th–18th century Chinese). The definite identification of carbon (in particular of soot or lampblack) as a pigment has posed problems where only a small sample can be taken, and the scanning electron microscope may be the instrument of choice in making such identification.

Carbon Manufacture. The history of Chinese ink indicates that the carbon was made in at least two ways—*viz.*, by burning wood, usually pine, or by burning oil. However, other fuels may also have been used.

It is helpful to know if the mode of manufacture can be distinguished for a given sample.

Despite the absence of char noted above, it would be surprising if there were not some reflection of the method of manufacture in the particle size distribution. An oil lamp has a steady flame of fairly constant size, while a wood fire has a variable, flickering flame. A naive, intuitive view would be that the wood fire can be expected to produce a broader range of particle sizes, whatever the mean and mode of the distribution might be. The constantly changing geometry of the flame will tend to allow some particles to escape before they have combusted very far, others perhaps before they are fully formed.

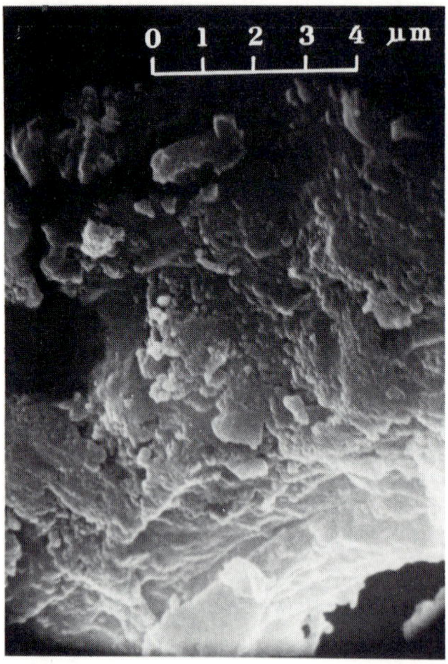

Figure 8. Scanning electron micrograph; ink on paper fiber from 14th century Chinese painting No. 31.1

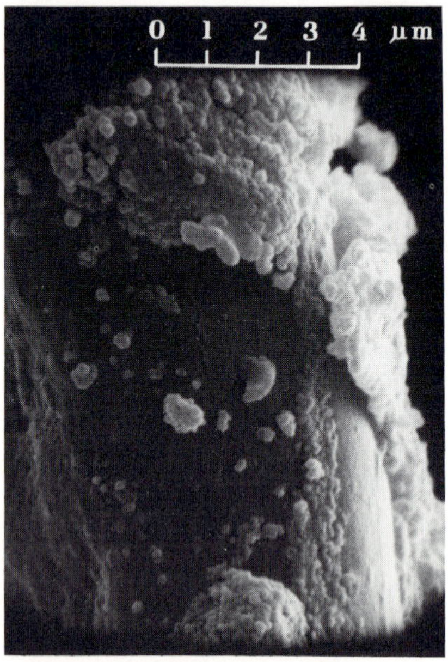

Figure 9. Scanning electron micrograph; ink on paper fiber from 17th–18th century Chinese painting No. 50.19

To some extent this view is borne out by observation. A modern ink known to have been made from lampblack (Figures 1 and 14) has a fairly uniform particle size, with a strong peak to the distribution. Figures 2 and 15 came from another modern ink made (according to details that came with it) from pine soot prepared in the traditional way. Despite inaccuracies in the plots noted above, the difference between them is large enough to afford some confirmation of expecta-

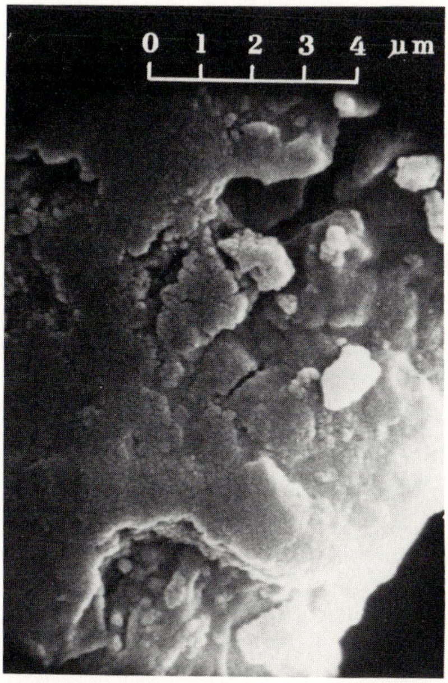

Figure 10. Scanning electron micrograph; ink on paper fiber from 17th century Chinese painting No. 61.11

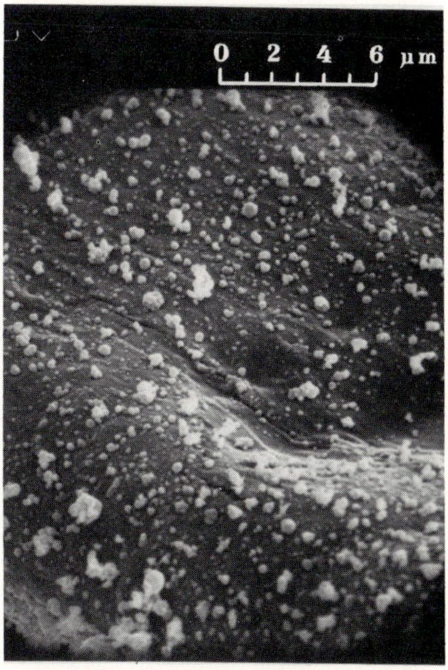

Figure 11. Scanning electron micrograph; ink on paper fiber from 18th century Chinese painting No. 62.5

tions, especially since it accords with the visual impression given by Figures 14 and 15.

The chimney soot sample (Figure 16) offers confirmatory evidence. Although the combustion particles are too ill-defined to permit a worthwhile particle size plot, a rough analysis suggested a particularly wide range of diameters, wider perhaps than those encountered in inks.

Another interesting case is that of the 14th century Chinese painting No. 31.1. The bluish tint of this ink suggests that it may be of the pine soot type (*see* below). The distribution (Figures 3, 6, 7, and 8) is narrow, but it falls in a range of considerably smaller diameters than lampblack ink. The plotted distribution is almost certainly truncated—the method used discriminated heavily against particle sizes in the < 0.05-μm class. In this sense the apparent narrowness of the distribution may be misleading. It is also tempting to suggest that a particularly rigorous soot fractionation may have occurred in this case, removing fairly large particles from the pigment used.

The overall impression is that pine soot has a variable particle size distribution but is quite broad in some cases at least. Present evidence is that lampblack has a distribution of roughly the kind in Figure 1. This type has turned up several times in Edo period Japanese paintings; such artists might have used lampblack ink which was the main type available in that late period.

Tint of Chinese Ink. It is unusual for a thin wash of ink on a white ground to appear absolutely neutral grey. Generally it has either a brownish or a bluish tint, and this fact is of interest in paintings. Brownish tints are the more common; in many actual paintings, of course, the distinction cannot be drawn because the support has darkened or because there is no sufficiently light wash of ink. The conventional view (*23, 28*) is that bluish ink is made from pine soot and brownish ink from lampblack. The question is therefore bound up with that of soot manufacture (*see* previous section). The 14th century Chinese painting (*cf.* Figures 3, 6, 7, and 8) shows a definite bluish tint, and this is the reason for suggesting that the ink may have been from pine soot. Apart from that, why do different samples of "black" ink differ slightly in their color? At least three reasons seem possible: (a) colored additives or secondary constituents, referred to in the section on history and manufacture; (b) light dispersion through scattering by particles; (c) differences of consumptive absorption [*i.e.*, absorption of radiant energy without scattering or re-emission (*29*)] by the different carbons.

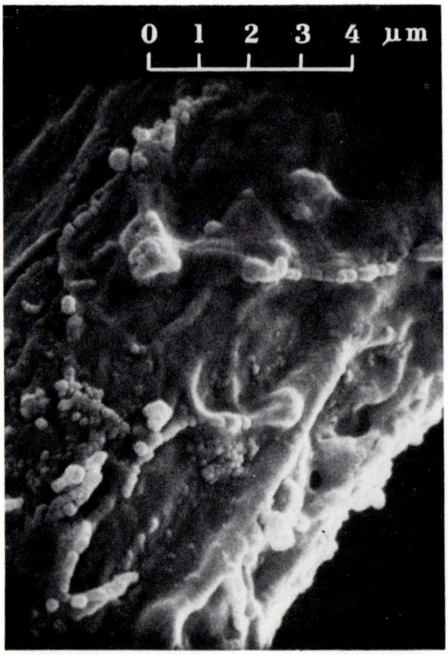

Figure 12. Scanning electron micrograph; ink on paper fiber from 13th–14th century Japanese painting No. 50.25

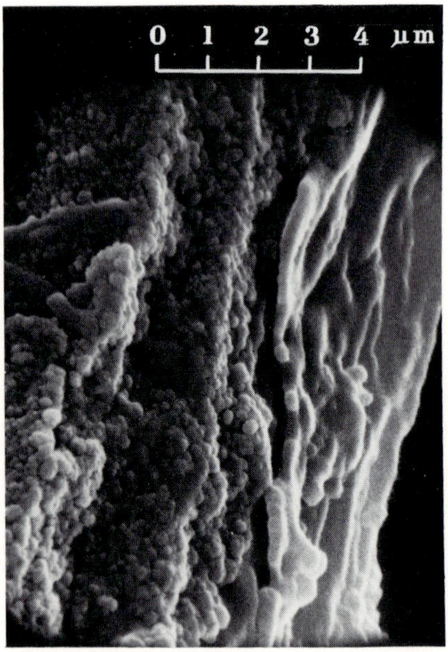

Figure 13. Scanning electron micrograph; ink on paper fiber from 18th century Korean painting No. 60.9

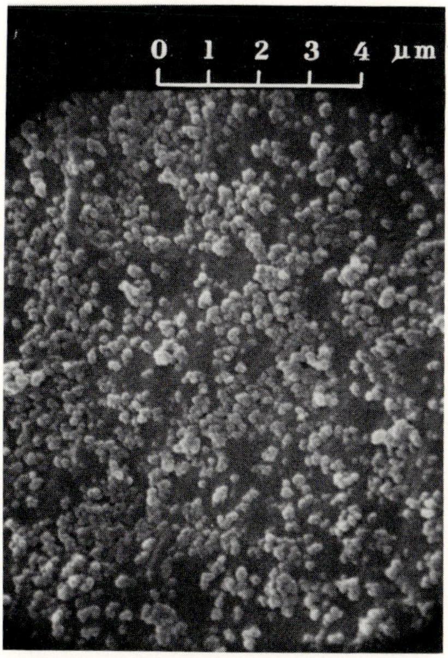

Figure 14. Scanning electron micrograph; paper fiber bearing modern lampblack ink

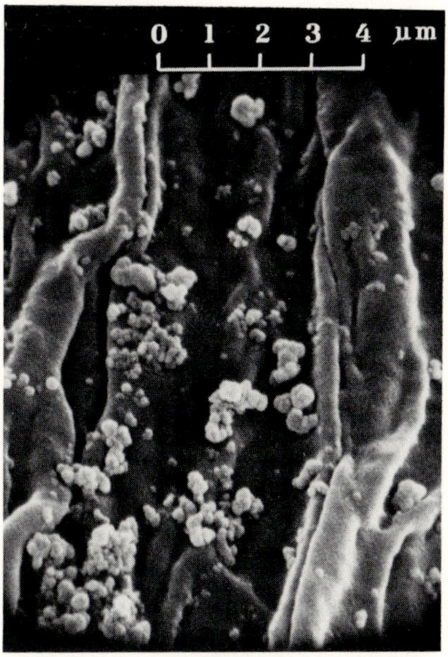

Figure 15. Scanning electron micrograph; paper fiber bearing modern pine soot ink having a bluish tone

ADDITIVES. No experimental evidence has been found for an additive affecting the tint of the ink. Exhaustive extraction of the modern bluish ink (*cf.* Figures 2 and 15) with solvents, including pyridine (which will dissolve indigo, for example) removed nothing. Combustion of a sample left a small residue, whose analysis gave no reason to suspect an inorganic pigment, and which was similar to the analysis of residue from a typical brownish ink. Further, carbons can display a blue tinge without added coloring matters. For example, a standard pigment sample from the National Bureau of Standards (No. 318) does so although it is certified to consist only of combustion carbon. Also, certain carbon blacks can have a blue undertone (*26*).

Clearly then, bluish inks can exist without coloring additives, and there is no experimental evidence that such additives are ever responsible. However, it is difficult to rule them out altogether; the historical literature refers to the use of bark from the pepper plant (*9*) and from *Fraxinus pubinervus* (*2*) to obtain a blue-tinged ink.

LIGHT DISPERSION. The idea of different tints being seen as a scattering spectrum as a result of different carbon particle sizes is intellectually attractive. A word on the general principles involved is in order.

Carbon particles in the size range we are considering can both absorb and scatter light. Light absorption and conversion to heat (consumptive absorption) are considered below. A further possibility is absorption and re-emission (fluorescence or phosphorescence) which are not explicitly considered here and seem unlikely. Scattering can also result in dispersion of the scattered light, and this is the possibility under consideration.

A general scattering theory for absorbing particles was developed by Mie (*30, 31*), and tables of values for the scattering functions have been computed (*32*). The earlier Rayleigh theory (*31*) is unsatisfactory for this case since it assumes non-absorbing particles of a smaller size range. The Mie theory is supported experimentally by the well known work on gold (*31*) and sulfur (*31, 33*) sols. It is mathematically complicated and involves functions of the complex refractive index, $m = n(1 - ik)$, where n is the real (or conventional) refractive index, and k is the absorption coefficient ($i^2 = -1$). It is partly the lack of knowledge of values for these parameters for a carbon particle coated with protein, and partly the difficulty associated with particles on an irregular surface (as opposed to a homogeneous sol) that has made it impractical to adapt

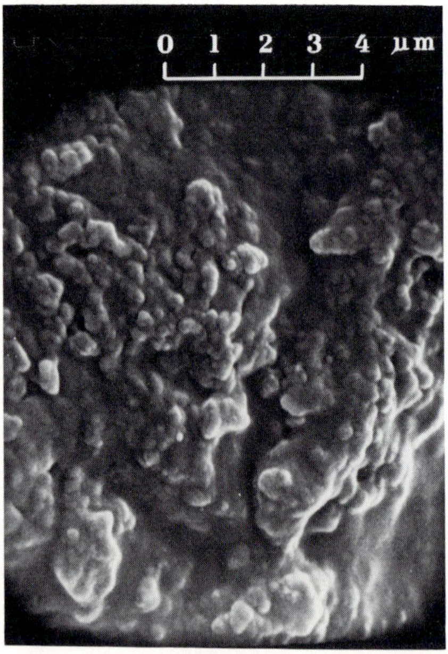

Figure 16. Scanning electron micrograph; household chimney soot derived from burning wood

the theory to the present case. Nevertheless, the hypothesis of a scattering dispersion of light has been rejected for the following reasons.

(a) Bluish inks are observed for quite different particle size distributions. For example, the modern bluish ink (Figures 2 and 15) has a distribution with a mode around 0.09–0.10 μm, with a long tail of larger particles. On the other hand, painting No. 31.1 (Figures 3, 6, 7, and 8), which has a similar tint, has a distribution with a mode around 0.05–0.06 μm and a very short tail. Although the painting from which Figure 11 was taken (No. 62.5) has a slight brownish tint, the distribution is generally similar to Figure 15, for the bluish ink.

(b) Work on gold and sulfur sols (*31, 33*) has shown that to obtain good scattering dispersion, reasonably uniform particle sizes are needed. This is certainly not the case with some of the inks studied.

(c) The propensity in some paintings for the carbon particles to be heavily aggregated casts doubt on how far they will function as independent scattering centers; it suggests that scattering might depend on such matters as the sizing of the painting support and how well the ink was dispersed. Such dependence appears not to have been noted.

CONSUMPTIVE ABSORPTION. There remain differences of consumptive absorption by the carbon, and it is reasonable that they should vary with the method of preparation. For example, carbons generally contain unpaired electrons (*34*), and this free radical character, at least during formation, is a function of position in the flame (*27*). That the free radical character should differ for soot and lampblack is therefore at least plausible, as is the dependence of light absorption on this character. Surface groupings on carbons can also differ (*35*), and this could affect light absorption. Thus, the various tints seen in thin washes of Chinese ink can be considered a result of differences in the consumptive light absorption by the carbon, and they may well reflect different methods of preparation.

Variability of Particle Size Distribution. How far is it possible to identify differences in inks more specific than those considered above? It is highly desirable, for example, to be able to associate a particle size distribution with a certain period or region or even with a particular school or artist. Although it is not possible to do so with the methods presented here, certain general points can be made.

Although carbon particle diameters are all of the same order, there is good evidence that even where the mean and mode of the distribution do not vary much, the overall shape may. The five ink histograms (Figures 1–5) show much error in the analysis, but the variance, skewness, and kurtosis of such distributions probably differ in a meaningful way. The three distributions from paintings vary from a heavy predominance of small particles (Figure 3) to a more equal distribution over a medium range of diameters (Figure 4) and to a long-tailed distribution with greater numbers of larger particles (Figure 5). Visual inspection of the scanning electron micrographs (Figures 6, 7, 8, 12, and 13) also shows that size differences exist. It is equally clear that visual impressions alone are misleading. Cursory inspection of Figure 13 suggests a predominance of larger particles. Analysis (Figure 5) confirms the greater numbers of larger particles but indicates that smaller ones may still be in the majority. The eye tends to ignore them in the micrograph itself.

How far particle size analysis will lead is uncertain. At an elementary level, the ability to confirm or deny the identity of ink in two areas of the same painting would itself be useful(—*e.g.*, in assessing the authenticity of a signature). This author's opinion is that the differentiation of lampblack inks is likely to be difficult or impossible because of the reproducible conditions of carbon manufacture. With pine soot inks (and perhaps others, if they exist) the position is more hopeful since fortuitous circumstances may show that ink made in one place differs from that made in another. With regard to the history of ink, this approach could be applied more often to earlier paintings, and its potential value is obvious.

Acknowledgment

The operation of the scanning electron microscope was in the hands of Walter R. Brown of the National Museum of Natural History. R. M. Organ and Harold Westley of the Smithsonian Institution supplied the spectrographic results for the ink ignition tests. H. P. Stern and T. Lawton (Director and Assistant Director of the Freer Gallery of Art) are thanked for advice on the selection of paintings, for permission to sample them, and for general discussions on the history of Chinese ink.

Literature Cited[a]

1. Lucas, A., "Ancient Egyptian Materials and Industries," 4th ed., rev. by J. R. Harris, Edward Arnold, London, 1962.
2. Laufer, B., in "Printing Ink, A History," Frank B. Wiborg, Ed., pp. 1-57, Harper and Brothers, New York and London, 1926.
3. Wang Chi-chen, *Metropol. Museum Studies* (1930) **3,** 114.
4. Franke, H., *Bayer. Akad. Wiss., Philosoph.-Histor. Klasse, Abhand.—N. F.* (1962) **54** (title: "Kulturgeschichtliches über die Chinesische Tusche").
5. Britton, R. S., Benedetti-Pichler, A. A., *Harvard J. Asiatic Studies* (1937) **2,** 1.
6. Benedetti-Pichler, A. A., *Ind. Eng. Chem., Anal. Ed.* (1937) **9,** (3), 149-152.
7. Creel, H. G., "Studies in Early Chinese Culture," p. 45, Waverley Press, Baltimore, 1937.
8. Chia Ssu-hsieh, "Ch'i-min yao-shu," (5th cent. A.D.); the relevant passage is translated into German in Ref. 4.

[a] Chinese and Japanese names are quoted surname first.

9. Jametel, M., "Chen-ki-souen. L'encre de Chine: son Histoire et sa Fabrication, d'après des Documents Chinois," Paris, 1882; translation into French of Shen Chi-sun, "Mo-fa chi-yao" (1398).
10. Tsien Tsuen-hsuin, "Written on Bamboo and Silk. The Beginnings of Chinese Books and Inscriptions," pp. 164-172, University of Chicago Press, 1962.
11. Julien, S., Champion, P., "Industries Anciennes et Modernes de l'Empire Chinois," pp. 129-140, Paris, 1869.
12. Julien, S., *Ann. Chim. Phys.* (1833) **53,** 308.
13. Sung Ying-hsing, "T'ien-kung K'ai-wu" (Chinese Technology in the 17th Century) translated by Sun E-tu Zen, Sun Shiou-chuan, pp. 285-287, Pennsylvania State University Press, 1966.
14. Li Ch'iao-p'ing, "The Chemical Arts of Old China," pp. 119-132, American Chemical Society, Washington, D.C., 1948.
15. Yü Fei-an, "Chung-kuo Hua Yen-se Ti Yen-chiu," (Research on Colors in Chinese Painting) Ch'ao-hua Art Publishing Co., Peking, 1955.
16. Mitsukuri Shinroku, *Nihon Gakujutsu Kyōkai Hōkoku* (1930) **6,** 147.
17. Hermans, P. H., in "Colloid Science," H. R. Kruyt, Ed., Vol. II, p. 562, Elsevier, New York, 1949.
18. Fieser, L. F., Fieser, M., "Organic Chemistry," 3rd ed., p. 452, Reinhold, New York, 1956.
19. Overbeek, J. Th. G., in "Colloid Science," H. R. Kruyt, Ed., Vol. I, p. 316, Elsevier, New York, 1952.
20. Terada Torahiko, Yamamoto Ryūzō, Watanabe Tetu, *Sci. Papers Inst. Phys. Chem. Res., Tokyo* (1934) **23,** 173.
21. Terada Torahiko, Yamamoto Ryūzō, Watanabe Tetu, *Sci. Papers Inst. Phys. Chem. Res., Tokyo* (1935) **27,** 75.
22. Yu Yü-hui, *Chemistry (China)* (1935) **2,** 779; *Chem. Abstr.* (1936) **30,** 2409.
23. Oguchi Hachirō, *Bull. Fac. Fine Arts, Tokyo Univ. Arts* (1969) (5), 27.
24. Miyasaka Kazuo, *Kōbunshi* (1966) **15,** 209.
25. Iijima Shigeru, "Kemboku Shingo," 2nd ed., Yūzan Kaku, Tokyo, 1943.
26. Gaydon, A. G., Wolfhard, H. G., "Flames. Their Structure, Radiation and Temperature," 2nd ed., Chapman and Hall, London, 1960.
27. Bradley, J. N., "Flame and Combustion Phenomena," pp. 116-119, Methuen, London, 1969.
28. Sugimura Yūzō, Nagai Toshio, "Bumbō Shihō—Sumi," Tankosha, Kyoto, 1972.
29. Kruyt, H. R., Ed., "Colloid Science," Vol. I, p. 34, Elsevier, New York, 1952.
30. Mie, G., *Ann. Phys.* [4] (1908) **25,** 377.
31. Jonker, G. H., in "Colloid Science," H. R. Kruyt, Ed., Vol. I, pp. 91-108, Elsevier, New York, 1952.
32. Lowan, A. N., "Scattering and Radiation from Circular Cylinders and Spheres. Tables of Amplitudes and Phase Angles," U.S. Navy Dept., Washington, D.C., 1946.
33. Johnson, I., LaMer, V. K., *J. Amer. Chem. Soc.* (1947) **69,** 1184.
34. Collins, R. L., Bell, M. D., Kraus, G., *J. Appl. Phys.* (1959) **30,** 56.
35. Caudle, J., Summer, K. G., Tye, F. L., *Proc. Conf. Ind. Carbons Graphite, 3rd, Imperial Coll. London, April 1970,* 168-171 (1971).

RECEIVED July 9, 1973.

12

Nuclear Magnetic Resonance Spectrometry in Archaeology

CURT W. BECK, CONSTANCE A. FELLOWS, and EDITH MACKENNAN

Vassar College, Poughkeepsie, N. Y. 12601

Proton magnetic resonance (PMR) spectrometry supplements optical spectroscopy and gas chromatography in the identification of organic archaeological materials. PMR establishes the average chain length and unsaturation of fatty acids. An oil sample of the sixth-fourth century B.C. is shown to consist largely of oleic acid. A solid fat of the third century A.D. contained principally myristic and palmitic acids. The absence of measurable glyceryl signals shows both samples were at least 95% hydrolyzed. PMR spectra also indicate that the pyrolysate of Baltic amber contains significantly less p-*cymene than that of modern pine resins, confirming previous evidence that abietane structures are only a minor component of the fossil resin.*

In little more than a decade, nuclear magnetic resonance (NMR) spectrometry has become an indispensable tool of organic chemistry, used as routinely as infrared spectroscopy which it supplements rather than replaces. Atomic nuclei with odd mass numbers such as ^{1}H, ^{13}C, ^{15}N, ^{17}O, ^{19}F, and ^{31}P and all atomic nuclei with even mass but odd atomic numbers such as ^{2}H, ^{10}B, and ^{14}N have magnetic moments and give NMR signals. Of these, carbon and hydrogen are found in practically all organic compounds, but while ^{13}C magnetic resonance (CMR) spectrometry requires very expensive equipment, the instrumentation of proton magnetic resonance (PMR) spectrometry is now within the financial reach of most laboratories.

In brief, a magnetic atomic nucleus in a magnetic field may be either lined up with or opposed to the external field, and the transition from one state to the other corresponds to an amount of energy which can be provided by electromagnetic radiation in the radiofrequency range.

NMR instruments use either constant magnetic fields and variable radiofrequencies or constant frequencies and variable magnetic fields, but the abscissa of NMR spectra is always measured in units of frequency (cps) in relation to an added reference compound such as tetramethylsilane (TMS) and divided by the radiofrequency of the instrument to yield dimensionless numbers on either the δ-scale, which sets the reference signal equal to zero, or on the τ-scale, which sets it equal to 10; thus $\delta + \tau = 10$. The spectra in this report use the δ scale and show the reference signal of TMS at $\delta = 0$.

The primary and secondary literature of NMR spectrometry is far too large to permit even a recapitulation here (*1*, *2*). We will attempt no more than a very brief summary of the observable parameters and their significance.

Chemical Shift. The position of the resonance signal of a proton (or of a group of identical protons) is an indication of the molecular environment of that proton and permits, *e.g.*, the distinction of aliphatic, allylic, olefinic, and aromatic hydrogens.

Integral. The area under an NMR signal (or a group of signals) is a measure of the number of hydrogen atoms which are responsible for that signal.

Spin-Spin Splitting. The splitting of a signal into two, three, four, or more peaks which show a binomial distribution pattern is an indication of the number of hydrogen atoms on neighboring carbon atoms which change the effective magnetic environment of the proton under observation by small but predictable amounts.

Coupling Constants. The distance between the individual peaks of a split NMR signal is a further indication of the molecular environment of the observed protons. For example, protons on neighboring saturated carbon atoms split each others' signals with coupling constants of the order of 6 cps; protons on neighboring olefinic carbon atoms ordinarily have coupling constants which are significantly greater.

Since the number of hydrogen atoms in an average organic molecule is quite large, most NMR spectra are fairly complex. Since many similar protons will give signals with nearly the same chemical shifts, resolution is often incomplete. The use of lanthanide shift reagents is a recent advance which goes far toward solving the latter problem: By coordinating a rare earth complex with the substrate molecule, protons are deshielded in inverse proportion to the third power of their distance from the coordination site, and thus the signals are spread over a wider range. Thus NMR spectrometry can yield information about the structure, and even the conformation, of pure compounds of considerable complexity. Mixtures, of course, confront the spectroscopist with special problems, but even here the NMR technique can often deal with the quantitative analysis of, for example, mixtures of isomers more effectively and effortlessly than can infrared spectroscopy. It is this aspect which makes NMR spectrometry particularly attractive to archaeological chemistry. The archaeological chemist is often confronted with very small amounts of complex mixtures. If the material is organic, the first step is to establish its general nature: is it a fat or oil, a pitch or tar, a resin or gum? While such broad distinctions as well as the final identification of the constituents can be made by various analytical methods, we have found NMR spectrometry uniquely useful in a number of recent studies.

Hydrolyzed Vegetable Oil

In the Newark Museum there is a small glass flask which is dated as sixth–fourth century B.C. with a general provenance of Syria. It was found to contain a dark brown liquid. The contents were removed by Robert H. Brill of the Corning Museum of Glass and sent to us for analysis.

An NMR spectrum of the substance dissolved in deuterochloroform (Figure 1) immediately showed it to be predominantly oleic acid (or a soluble metal oleate), probably with some admixture of saturated fatty acids. The signals can be assigned as follows. The triplet at 0.8 δ arises from three protons of a terminal methyl group attached to a methylene group. The large, broad singlet at 1.1 δ represents 22 protons on methylene groups which are not deshielded by neighboring sp^2 carbon atoms and which are so similar to one another that no measurable splitting occurs. The ill-resolved peaks between 1.6 and 3.0 δ are caused by methylene groups which are deshielded by neighboring sp^2 carbon atoms. The carboxyl group accounts for one such methylene group. Since the

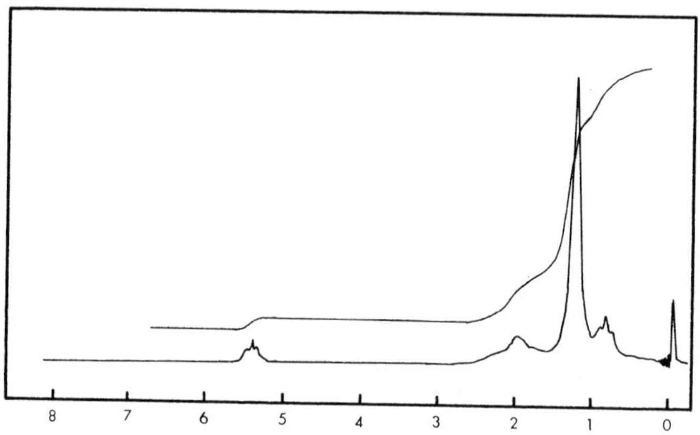

Figure 1. Oil from Syrian glass bottle, sixth–fourth century B.C.

integral in this area indicates less than six protons, there can be no more than two other deshielded methylene groups; hence there is at most one double bond per molecule. The two vinylic protons of this double bond produce the signal at about 5.4 δ. The integral of this signal corresponds to only 1.3 protons. The deficiency of deshielded methylene and of vinylic protons indicates an admixture of saturated fatty acids.

Oleic acid is the principal fatty acid of olive oil (83%), which also contains smaller amounts of the saturated palmitic (6%) and stearic (4%) and of the doubly unsaturated linoleic acid (7%). The NMR spectrum of the archaeological sample was closely matched by spectra of commercial oleic acid (Figure 2) as well as by mixtures of the acid with its sodium and potassium salts (Figure 3).

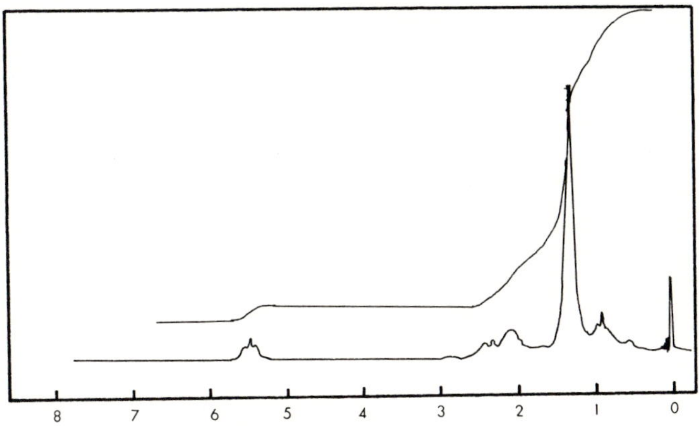

Figure 2. Oleic acid

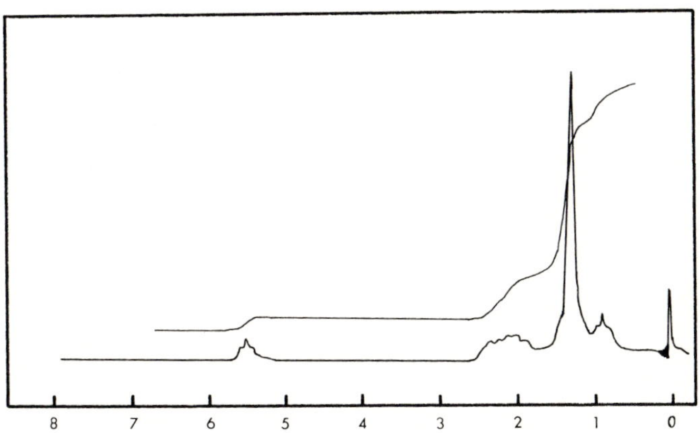

Figure 3. Oleic acid containing 21% sodium oleate

In spite of its considerable age, one might suspect that the archaeological material could have retained some unsaponified olive oil, and we therefore attempted to establish the limits of detection of the unsaponified oil in the presence of free fatty acids or salts. The NMR spectrum (Figure 4) of the simplest triglyceride, triacetin (the ester of glycerol with acetic acid) shows the resonance of the four protons of the methylene groups of glycerol as a multiplet at about 4.3 δ (the multiplicity of the signal is caused by the magnetic non-equivalence of the two protons in each methylene group). The single tertiary proton of the glyceryl group appears essentially as a quintet at about 5.3 δ. The latter position is the same as that of the vinyl protons in oleic acid, and the two signals cannot be used to distinglish the glyceryl ester and the free fatty acid on our instrument. The NMR spectrum of fresh olive oil (Figure 5) confirms this; the vinyl protons of the unsaturated fatty acid and the tertiary pro-

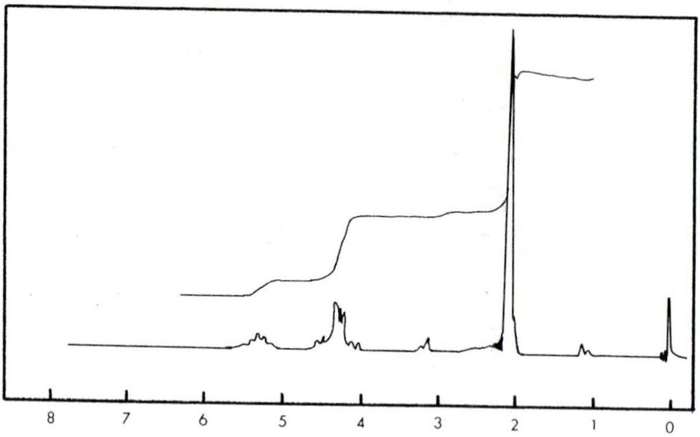

Figure 4. Triacetin

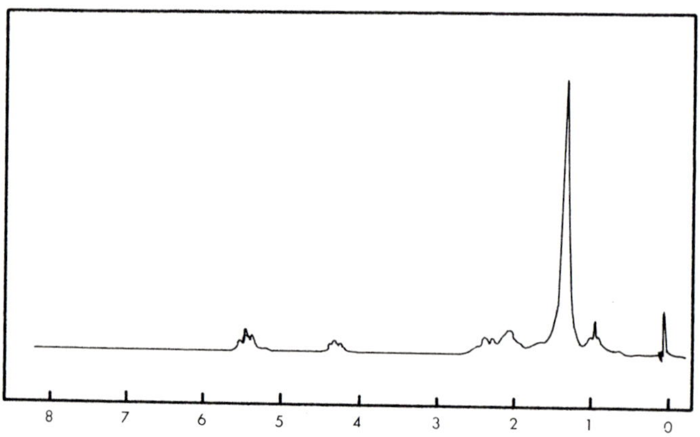

Figure 5. Italian olive oil

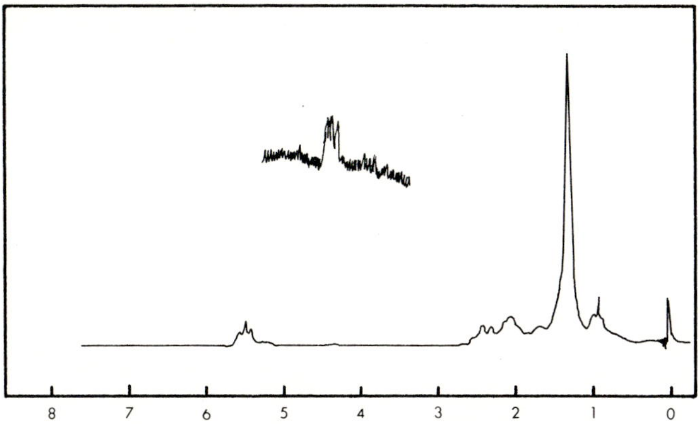

Figure 6. Oleic acid containing 10% olive oil

ton of the glycerol moiety coincide, but the signal of four methylene protons is clearly visible. In a synthetic mixture of oleic acid containing 10% olive oil (corresponding to 0.5% free glycerol) (Figure 6), the glyceryl protons at 4.3 δ are barely visible as a disturbance in the baseline at low amplitudes, but after amplification they are clearly identifiable in spite of the large amount of noise which is the price of electronic amplification. However, 5% olive oil (corresponding to 0.25% free glycerol) can no longer be detected by our Varian T-60 instrument. The limits of detection could be lowered substantially by using a more sensitive spectrometer and a time-averaging computer which stores many repeated spectral runs and prints out their sum; the signals grow in proportion to the number of scans superimposed while the noise increases only in proportion to the square root of that number, so that 100 repeated runs will improve the signal-to-noise ratio by a factor of 10.

The identification of the ancient oil sample was confirmed by other analytical techniques, including infrared spectroscopy, titration, and gas chromatography. A full account will be published elsewhere; our purpose here is to show how much useful information could be gained from NMR spectrometry alone.

Hydrolyzed Animal Fat

Another application concerns the solid residue in a pilgrim flask of the third century A.D. from the Rhineland, which was also referred to us by Robert H. Brill. About 70% of the material was soluble in carbon tetrachloride, and its NMR spectrum (Figure 7) immediately identified it as a mixture of saturated fatty acids (and their soluble salts) with an average chain length of 14 to 16 carbon atoms, corresponding to myristic and palmitic acids. This average molecular weight was obtained by

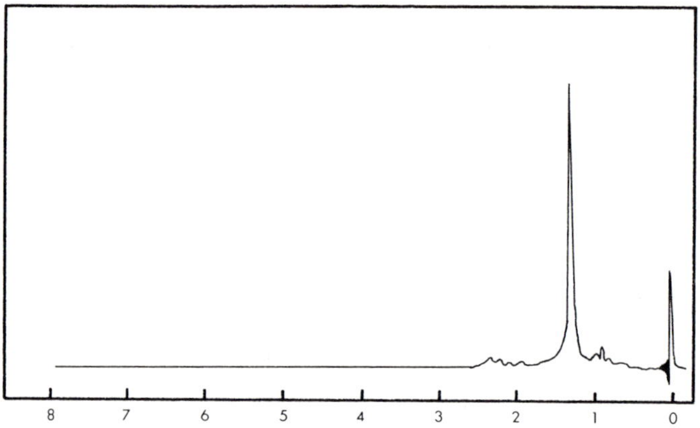

Figure 7. Fat from pilgrim flask, third century A.D.

comparing the integral of the terminal methyl group (which is known to contain three protons) and the integral of the α-methylene group (which is known to contain two protons) with the integral of the large singlet produced by essentially identical normal methylene groups of the saturated straight chain. This method cannot be highly accurate because the signals used are too close to permit precise measurement of their integrals. Furthermore the answer obtained is necessarily an average chain length of a mixture; the actual composition must be established by gas chromatography of the methyl esters. In the present case the latter technique showed that myristic acid is indeed the principal fatty acid present. However, NMR spectrometry can be used very effectively to determine the chain length, and hence identify the acid, of a single pure component after chromatographic separation; the methyl ester group introduces a new signal at about 3.7 δ which is well separated from all other signals and can therefore be integrated precisely. It is known to contain three protons. A comparison of its integral with the integral of all the other signals combined is an excellent measure of chain length as shown by the example of methyl palmitate (Figure 8) which shows an error of less than one proton.

The Structure of Amber

Another promising application of NMR spectrometry to archaeological problems deals with fossil resins. These terpenoid polymers, the best known of which is amber, have played a large role in trade relations during prehistoric times since at least the neolithic era, and the determination of their geographic and botanical origin can contribute to our knowledge of this early commerce.

Palaeobotanists have assigned the source tree of the Baltic amber of northern Europe to an extinct species of pine, *Pinus succinifera,* on purely morphological grounds but have been unable to relate this tree to any living pine species (*3*). This assignment has led some investigators to the hypothesis that Baltic amber may be a derivative of the principal resin acid of most common modern pines—abietic acid (*4*). Chemical and infrared spectroscopic evidence raises some serious difficulties with this view (*5*) but without being able to resolve the question. One crucial point concerns the presence of isopropyl groups in Baltic amber. Such groups are present in abietic acid and in all the polymeric derivatives of abietic acid which have been considered as constituents of fossil amber, but they are absent in the principal resin acids of many other possible source trees, including not only other conifers but also North American and Asiatic pines which might well have flourished in the mild climate of the European Tertiary period.

Unfortunately, fossil resins are largely insoluble because of their polymeric character, and a dissolved sample is a prerequisite for a well resolved NMR spectrum. We have therefore carried out comparative NMR studies on the soluble pyrolysates of amber and of pine resins.

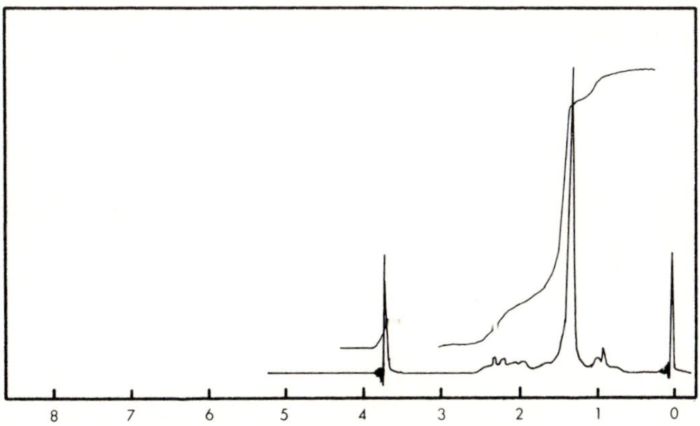

Figure 8. Methyl palmitate

The pyrolysis of pine rosin yields pine oil which has long been known (*6*) to contain large amounts of *p*-cymene (*p*-isopropyltoluene, bp, 177°C). We have prepared pine oil by dry distillation of commercial rosin and fractionated it at atmospheric pressure into 10° cuts. The NMR spectrum of the fraction boiling at 170°–180°C (Figure 9) shows that *p*-cymene is indeed the major constituent present. The NMR spectrum of pure *p*-cymene has been reported (*7*), and its signals are prominent in the spectrum of pine oil. The doublet at 1.22 δ, with a coupling constant of 7 cps, is caused by the two identical methyl groups of the isopropyl structure, split by the neighboring tertiary proton. This latter proton accounts for the multiplet (theoretically a septet) at 2.85 δ. The singlet of the toluene–methyl group appears at 2.30 δ in pure *p*-cymene. The spectrum of the pine oil fraction shows two singlets, at 2.28 and 2.29 δ, respectively. Either may be the *p*-cymene methyl group; the other is most likely caused by an admixture of the isomer *m*-cymene. The presence of other isomers is also evident from the multiplicity of the aromatic ring protons. While pure *p*-cymene gives a sharp singlet at 7.08 δ, pine oil shows splitting of the benzenoid protons which again may be accounted for by the presence of the meta isomer. Both isomers would be expected from the pyrolytic cleavage of an abietic acid molecule. The same signals of *p*-cymene are also present, although at lower intensities, in the NMR spectra of the pine oil fractions boiling at 160°–170°C, 180°–190°C, and 190°–200°C. Thus NMR offers a simple means of establishing the presence of isopropyl groups in resins of the abietic acid type.

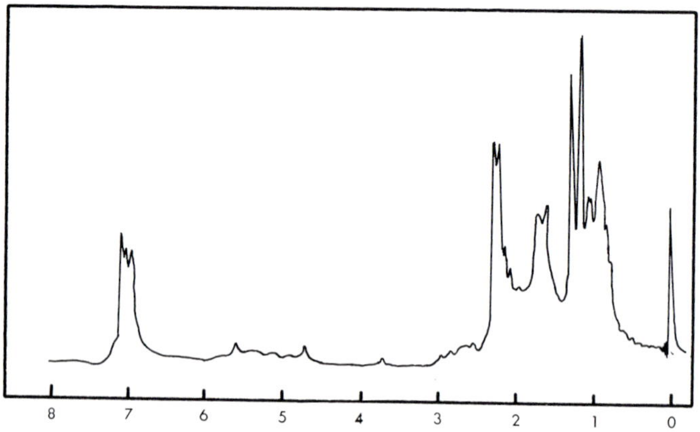

Figure 9. Pine oil fraction, boiling point 170°–180°C

Oil of amber obtained by destructive distillation of Baltic amber yielded only a very small fraction boiling in the range of 160°–180°C. The NMR spectrum of this fraction (Figure 10) lacks the prominent isopropyl doublet. In its approximate place there appears a characteristic signal of three evenly spaced peaks. It cannot be a proper triplet because the distance between the peaks is only about 3.5 cps and because the intensity of the peaks is not in the required 1:2:1 ratio. It seems therefore likely that this triple signal is merely an accidental superimposition of a doublet having a normal coupling constant of 7 cps with another quite unrelated singlet. If so, the doublet may still be the result of an isopropyl group, but the weights of the fractions and their spectra show

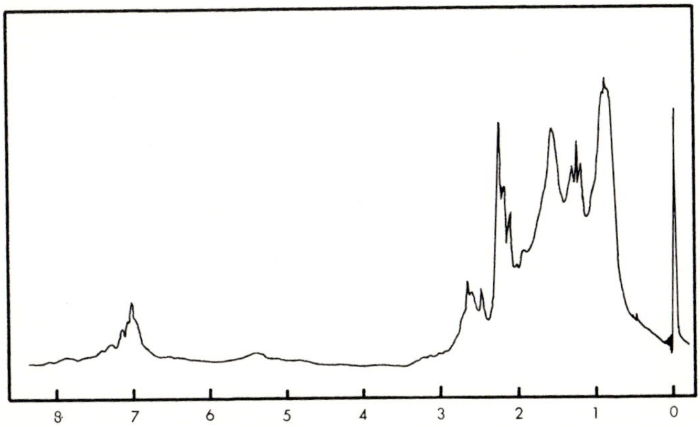

Figure 10. Oil of amber fraction, boiling point 160°–180°C

that amber yields significantly less *p*-cymene than does resin. Current work on the isolation of individual alkylbenzenes from resin pyrolysates followed by NMR analysis of the pure components may provide partial structures for the constituents of Baltic amber. Even now, however, NMR spectroscopy has confirmed that there are essential structural differences between Baltic amber and the abietic acid resins of the common species of living pines. These few random examples illustrate the potential contributions which NMR spectrometry is able to make to the archaeological chemistry of organic materials.

Literature Cited

1. Roberts, J. D., "Nuclear Magnetic Resonance: Applications to Organic Chemistry," McGraw-Hill, New York, 1959.
2. Bible, R. H., "Interpretation of NMR Spectra; An Empirical Approach," Plenum Press, New York, 1965.
3. Schubert, K., *Beiheft z. Geol. Jahrb.* (1961) 45.
4. Rottländer, R. C. A., *Archaeometry* (1970) **12**, 35.
5. Beck, C. W., *Naturwissenschaften* (1972) **59**, 294.
6. Kelbe, W., *Ann.* (1882) **210**, 1.
7. Varian Associates, "High-Resolution NMR Spectra Catalog," 1962, spectrum 268.

RECEIVED July 9, 1973.

13

Chemical Investigations of Ancient Near Eastern Archaeological Ivory Artifacts

N. S. BAER

Conservation Center, Institute of Fine Arts, New York University,
1 East 78th St., New York, N. Y. 10021

N. INDICTOR

Chemistry Department, Brooklyn College, City University of New York,
Brooklyn, N. Y. 11210

The composition of buried bone changes with time. Hydroxyapatite, the major inorganic component, undergoes very slow exchange with ionic species from groundwaters. The nitrogen content decreases as the collagenous material is removed and replaced, in part, by inorganic salts. The rate of nitrogen loss is generally much greater than the rate of the exchange reaction. The results of carbon, hydrogen, nitrogen, and ash analyses for samples of ivory excavated in the Ancient Near East reveal a similar compositional variation. In excavated specimens of known provenance (Hasanlu, Northwest Iran; Nimrud, Iraq; and Acem Hüyük, Anatolia) the nitrogen content was $\leq 0.5\%$ (except for black ivories from Hasanlu) whereas the nitrogen content of modern elephant ivory is $> 5.0\%$. The percent ash, as combustion residue, was $> 80\%$ for excavated ivory compared with ca. *55% for modern elephant ivory. The generality of these results for many samples has suggested the use of these data to develop criteria for the chemical examination of artifacts of doubtful provenance.*

Artifacts and works of art have been carved from bone and ivory from prehistoric times. The importance of these objects to art historical scholarship is well documented. In recent years, many ivory artifacts have been recovered in excavations in the Ancient Near East from such sites as Nimrud, Hasanlu, Acem Hüyük, and Khorsabad. These carvings have attracted considerable art historical and archaeological attention (*1, 2, 3, 4, 5*). The excavated ivories range from white through brown, grey, blue and black. The mechanism of color formation by the action of high temperature has been described (*6, 7*). In this paper the question of authenticity is considered for ivories of unknown provenance said to be from the Ancient Near East.

In 1958 the Metropolitan Museum of Art purchased a group of ivories said to be from Khorsabad, ancient Dûr Sarrukin, the unfinished capital built (713–707 B.C.) for King Sargon (Sarrukin) of Assyria (*2, 8*). Considerable difference of opinion exists among art historians and archeaeologists as to the assignment of these ivories to Khorsabad. There is also some question of their authenticity as ancient Assyrian ivories.

Chemical methods of dating have been applied to prehistoric bone specimens with varying degrees of success (*9, 10*). The most notable example is the case of the Piltdown forgeries in which many analytical methods were used (*11, 12*). The fluoride exchange method, originally proposed in the last century (*13, 14*) was reintroduced for this problem by Oakley and has since found wide application for excavated bone. The method depends on the gradual accumulation by the specimen of fluoride from groundwaters. Other systematic changes such as the variation of uranium, carbonate, nitrogen, and phosphorus content have been used in conjunction with fluoride data. The carbon-14 method has been used with some success under special circumstances (*15, 16*). For the past 25 years, chemical analysis of bone for dating purposes has focused on changes in the proteinaceous material.

In this work, we have studied the change in collagen content, by elemental analyses, for buried ivory obtained from archaeological excavations. The sites mentioned in this study are given in Table I. The results of the analyses for excavated specimens are compared with those for ivories of unknown provenance.

Experimental

Sample Preparation. Samples from objects were obtained by drilling deeply (*ca.* 2 cm) into the object, discarding the initial superficial material. The amount of useable sample obtained was generally 2–3 mg per

Table I. Locations of Ancient Near Eastern Archaeological Sites and Period Attributions of Specimens Examined

Site	*Period*	*Description*
Hasanlu	9th century B.C.	Northwest Iran, Period IV
Acem Hüyük	19th-18th century B.C.	Anatolia
Nimrud	9th-8th century B.C.	Iraq
Khorsabad	8th century B.C.	Ancient Dûr Sarrukin, Iraq

drilling. When expendable fragments were available, their surface was cleaned by mechanical scraping of the surface with a scalpel.

Analyses. Analyses of ash, carbon, and hydrogen by combustion, and nitrogen by Dumas or micro-Kjeldahl methods (*17*) were obtained from Schwarzkopf Microanalytical Laboratory, Woodside, N. Y. 11377.

Analytical Errors. Errors associated with these data arise from uncertainties in the analytical methods and from impurities introduced into or present in the ivory matrix. These added materials are either inorganic salts from groundwater replacing collagenous matter during burial or organic material used in the conservation laboratory to impregnate friable specimens.

The uncertainties for the analytical methods are as follows: ash by combustion residue, $\pm 0.3\%$ absolute; C, ± 0.05 mg C; H, $\pm$ 0.3% absolute; and N, by Dumas, ± 5 μl or by micro-Kjeldahl, $\pm$ 1 μg (*18*). The large uncertainties in the measurements of carbon and hydrogen limit the usefulness of data for these elements for very small samples.

Errors associated with inorganic salts arise from the presence of added carbonate and nitrate. The results of leaching experiments with HCl to eliminate carbonate indicate the presence of negligible additional carbonate (*6, 19*). The use of NO_3^- insensitive methods for nitrogen demonstrated the absence of measurable nitrates.

Methods for strengthening friable ivory artifacts are described by Plenderleith and Werner (*20*). The materials used will lead to high carbon and high hydrogen values generally and high nitrogen values only when nitrogen is a constituent of the impregnating material—*e.g.*, soluble nylon, nitrocellulose, etc. Discarding the superficial material tends to eliminate this source of error except in vacuum impregnation. When the treatment history of an artifact indicates vacuum impregnation, solvents may be used to leach out the extraneous matter.

Discussion

Related Materials. Table II gives the results of elemental analyses for bone, ivory, and related specimens. The mastadon, mammoth, and ancient walrus sample analyses indicated retained proteinaceous matter in composition equivalent to that for modern elephant ivory. This suggests that these materials would not provide a ready source of "aged" ivory for a modern carver seeking to duplicate the texture and composition of ivory buried for well over 2,000 years.

Acem Hüyük. The data in Table III for these, the oldest artifacts examined, reveal a consistent loss of organic matter with no detectable nitrogen. This would indicate a complete loss of proteinaceous matter with some residual free carbon. As previously reported (*6*), the presence of blue, black, or grey coloration in ivory has been attributed to the thermal decomposition of the collagen with a consequent deposition of free carbon. Some loosely held water was occasionally present. Leaching experiments (samples marked Acem Hüyük IV) demonstrate the possible presence of only a small amount of carbonate.

Table II. Carbon, Hydrogen, Nitrogen, and Ash Analyses for Specimens of Ivory and Related Materials

Specimen	*% Ash*	*% C*	*% N*[a]
Elephant ivory (Africa)	53.32	16.25	5.52
Mammoth tusk (N.E. Siberia)[b]	54.24	15.64	5.37
Mastadon[b]	56.04	16.85	5.78
Sperm whale tooth	66.02	12.28	4.23
Whale bone (rib)	71.62	11.85	3.03
Hippopotamus tooth	60.36	12.64	4.25
Boar tooth (Aitape, New Guinea)	65.57	12.25	3.86
Ancient walrus tusk (Alaska)[c]	60.46	15.16	4.75
Beefbone[d]	68.68	10.94	3.52

[a] All analyses by Dumas method.
[b] Pleistocene Epoch, courtesy The American Museum of Natural History.
[c] Fossilized specimen, courtesy The American Museum of Natural History.
[d] Collagenous material extracted with boiling water for 2 hrs.

Nimrud. The samples analyzed from Nimrud were all white (Table IV). In general, the ash content was approximately 85% while the nitrogen content varied from 0–0.4%. The general loss of organic matter is again observed, though a small amount of nitrogenous material has survived.

Hasanlu. These specimens fall into two distinct classes: black objects and objects which are some shade of grey (Table V). The grey objects have compositions not unlike the specimens from Nimrud (Table IV). However, the black objects demonstrate a significantly higher concentration of proteinaceous material. The well preserved black object, Hasanlu

Table III. Analyses and Color of Ancient Ivory Specimens from Acem Hüyük, Anatolia (19th-18th Century B.C.)

Specimen	*Color*	*% Ash*	*% C*	*% H*	*% N*[a]
Acem Hüyük I	black-grey	78.43	2.65	0.76	0
Acem Hüyük II	blue-grey	92.60	2.12	0.41	0
Acem Hüyük III	grey-white	97.34	0.41	0.23	0
Acem Hüyük IV[b]	white	98.86	0.38	0.09	0
Acem Hüyük IV[c]	white	98.81	0.33	0.05	nd[d]
Acem Hüyük IV[e]	white	nd	0.22	nd	nd

[a] All anlyses by Dumas method.
[b] 20-mg sample.
[c] 70-mg sample.
[d] No data.
[e] After treatment with HCl to remove carbonates.

V, described elsewhere (*4*), represents our only example of an excavated ivory with a nitrogen content greater than 1%. The presence of free carbon seems to have protected partially the collagen from the action of groundwater and bacteria. This observation is supported by the gen-

Table IV. Analyses and Color of Ancient Ivory Specimens from Nimrud, Iraq (9th-8th Century B.C.)

Specimen	*Color*	*% Ash*	*% C*	*% H*	*% N* [a]
Nimrud I	white	85.6	4.33	0.86	0
Nimrud II	white	82.2	4.39	1.21	0
Nimrud III	white	82.3	5.45	1.18	0.39
Nimrud III [b]	white	nd [c]	nd	nd	0.19 (K)
Nimrud IV [d]	white	83.62	5.12	1.04	0.25
Nimrud V	white	87.16	2.52	0.87	0.36
Nimrud V	white	nd	nd	nd	0.13

[a] All analyses by Dumas method except when marked (K) for Kjeldahl.
[b] 2.5-mg sample.
[c] No data.
[d] Metropolitan Museum of Art accession number 59.107.9, fragment.

erally better preserved state of excavated black ivory when compared with ivories of other color (*21*).

Khorsabad. The "Khorsabad" ivory specimens examined are described in Table VI. Figures 1 and 2 show two of the objects—a winged figure with bucket and an open-work griffin. Analytical data for their elemental analyses are presented in Table VII. In general, approximately

Table V. Analyses and Color of Ancient Ivory Specimens from Hasanlu, Northwest Iran, Period IV (9th Century B.C.)

Specimen	*Color*	*% Ash*	*% C*	*% H*	*% N* [a]
Hasanlu I	grey	87.15	2.36	0.65	0.53
Hasanlu II	white-grey	84.35	3.11	0.92	0.52
Hasanlu III	grey	87.55	2.90	0.63	0.54
Hasanlu IV	black	87.48	4.40	0.47	0.95
Hasanlu V [b]	black	77.69	8.52	1.18	1.50 (K)

[a] All analyses by Dumas method except when marked (K) for Kjeldahl.
[b] Metropolitan Museum of Art accession number 65.163.21.

2 mg of sample were used in making the analyses. Specimen III, a fragment, permitted a greater sample size. The data are clearly comparable with those for the excavated specimens from Hasanlu and Nimrud. The high ash values and generally low nitrogen analyses demonstrate a considerable loss of proteinaceous material.

In Table VIII, the composite values for all analyses, by group, are presented together with the analyses for modern elephant ivory and for an "Assyrian" ivory of unknown provenance in a private collection. The results of the elemental analyses for the Khorsabad ivories are consistent with those for the excavated specimens. Samples from the Assyrian ivory from the private collection, however, yield analyses which are not con-

Table VI. Designations of "Khorsabad" Ivory Specimens (Provenance Unknown)

Specimen	*Accession Number*[a]	*Description*
Khorsabad I	58.122.7	winged figure with bucket
Khorsabad II	58.122.1-11	drilling from fragment
Khorsabad III	58.122.1-11	chip from fragment
Khorsabad IV	58.122.10	griffin, striding open work
Khorsabad V	58.122.11	lion, carved in the round

[a] Metropolitan Museum of Art.

—Courtesy The Metropolitan Museum of Art, Fletcher Fund

Figure 1. Open work ivory plaque, standing winged figure with bucket. Height, 15.5 cm; width, 7.9 cm. Metropolitan Museum of Art accession number 58.122.7. Reputedly from Khorsabad.

—Courtesy The Metropolitan Museum of Art, Fletcher Fund

Figure 2. Open work striding griffin, decorated on both sides. Height, 5.4 cm; width, 10.3 cm. Metropolitan Museum of Art accession number 58.122.10. Reputedly from Khorsabad.

Table VII. Analyses and Color of "Khorsabad" Ivory Specimens (Provenance Unknown)

Specimen	*Color*	*% Ash*	*% C*	*% H*	*% N*[a]
Khorsabad I	white	90.4	4.1	1.5	0.5 (K)
Khorsabad II	white	nd[b]	nd	nd	0.4 (K)
Khorsabad III	white	84.2	5.2	1.1	0.2
Khorsabad IV	white	83.3	5.6	0 (?)	0.5 (K)
Khorsabad V	white	81.8	2.9	1.4	0.4 (K)

[a] Analysis by Dumas method where not marked (K) for Kjeldahl.
[b] No data.

Table VIII. Comparison of Composite[a] Values for Analyses of Excavated Ancient Near Eastern Ivory Specimens with Specimens of Unknown Provenance

Specimen	*Date*	*Color*	*% Ash*	*% C*	*% H*	*% N*
Elephant Ivory	modern	white	53.32 (3)	16.25 (3)	3.52 (3)	5.52 (3)
Nimrud	9th-8th century B.C.	white	84.2 (5)	4.36 (5)	1.03 (5)	0.19 (7)
Hasanlu	9th century B.C.	grey	86.4 (3)	2.78 (3)	0.73 (3)	0.53 (3)
Hasanlu	9th century B.C.	black	82.6 (2)	6.46 (2)	0.83 (2)	1.23 (2)
Acem Hüyük	19th-18th century B.C.	white	91.94 (6)	1.02 (6)	0.31 (5)	0 (4)
Khorsabad[b]	unknown	white	84.9 (4)	4.4 (4)	1.0 (4)	0.4 (5)
Assyrian[c]	unknown	white	60.7 (2)	10.83 (2)	2.77 (2)	3.65 (2)

[a] Composite value refers to the average of all sample analyses. Number of analyses appears in parentheses.
[b] *See* Table VI for description of specimens.
[c] Object of uncertain provenance in private collection.

sistent with those for Ancient Near Eastern ivories, thus far analyzed, of known provenance.

Conclusions

Elemental Analyses. The data for samples obtained from ivories excavated at three Ancient Near Eastern sites (Nimrud, Acem Hüyük, Hasanlu) demonstrate a uniform behavior similar to that of buried bone. The fundamental change is the decrease in nitrogen content as this collagenous material is removed and replaced, in part, by inorganic salts. In this study, the nitrogen content was $\leq$ 0.5%, except for the black ivories from Hasanlu whereas the nitrogen content for modern elephant ivory is $>$ 5.0%. The percent ash, as combustion residue, was $>$ 80% for excavated ivory *vs. ca.* 55% for modern elephant ivory. The limited scope of this study must be expanded to include additional sites, more black ivories, and a greater sample universe before such analytical data may be applied with confidence as criteria for the authentication of ivories of unknown provenance. Further work is also needed to characterize more thoroughly the changes in composition accompanying both natural and artificial aging of ivory.

Additional Methods of Analysis. Several methods of analysis, other than those reported above, have been used with varying success to the relative dating and, in some cases, to the absolute dating of bone. The limited amounts of sample available for analysis, typically on the order of a few milligrams, currently preclude radiocarbon and thermoluminescence dating. Two other exchange processes observed in buried bone; F^- exchange with OH^- to form fluorapatite, and the exchange of uranium for calcium in the apatite lattice, should be considered. Though these processes are far more characteristic of the specific archeological site than is collagen loss, they should provide valuable information supplementing the changes in the organic components of ivory. Preliminary fluoride analyses on excavated ivories indicate considerable promise for some archaeological sites while others appear to be insensitive (*22*).

The critical problem of sample size and the limitations imposed thereby on the accuracy of elemental analyses, especially those for carbon and hydrogen, suggest the application of chromatographic techniques. Amino acid analysis may provide a superior method of assessing collagen loss with additional information regarding the selective decomposition processes, sensitive to local soil conditions.

Ivories of Unknown Provenance. The data for the Khorsabad ivories are consistent with the suggested age. These data do not, however, eliminate the possibility of artificial methods for simulating the patina and texture, generally acquired by long-term burial, by chemical extraction of the collagen of modern ivory. Our experience suggests that such treatment effecting only limited removal of the collagen is usually accompanied by severe loss of mechanical strength. The data for the Assyrian ivory cited in Table VIII are not consistent with those for any of the excavated specimens.

Related Problems. The first question to be resolved in examining an Ancient Near Eastern ivory object of unknown provenance is its antiquity. Perhaps equally significant is the question of assignment to a specific archaeological site. This problem, as yet unexplored, may ultimately depend on the study of trace element distributions for its technical resolution.

At this time the most powerful technique available to study these problems is the art historian's stylistic analysis. The cooperation of archeologists and art historians will be a prerequisite for progress in the technical analysis of ivories.

Acknowledgment

We thank Vaughn E. Crawford, Oscar White Muscarella, and Prudence Oliver Harper of the Metropolitan Museum of Art for samples of ancient ivory, permission to sample the Khorsabad ivories, and for stimulating discussions.

Literature Cited

1. Barnett, R., "A Catalogue of the Nimrud Ivories, with other Examples of Ancient Near Eastern Ivories in the British Museum," British Museum, London, 1957.
2. Wilkinson, C. K., "The First Millennium B. C.," *Metrop. Museum Art Bull.* (1960) **18**, 261-268.
3. Mallowan, M. E. L., "Nimrud and its Remains," Dodd, Mead, New York, 1966.
4. Muscarella, O. W., "Hasanlu 1964," *Metrop. Museum Art Bull.* (1966) **25**, 121-135.
5. Mallowan, M., Davies, L. G., "Ivories in the Assyrian Style," The British School of Archaeology in Iraq, London, 1970.
6. Baer, N. S., Indictor, N., Frantz, J. H., Appelbaum, B., "The Effect of High Temperature on Ivory," Studies in Conservation (1971) **16**, 1-8.
7. Baer, N. S., Appelbaum, B., Indictor, N., "The Effect of Long-Term Heating on Ivory," *Int. Inst. Conserv., Amer. Group Bull.* (1971) **12**, 55-59.
8. Langer, W. L., Ed., "An Encyclopedia of World History," rev. 3rd ed., pp. 25-38, Houghton-Mifflin, Boston, 1952.
9. Groff, D. W., "Gas Chromatography Methods for Bone Fluorine and Nitrogen Components," in "Science and Archaeology," Chap. 20, MIT Press, Boston, 1971.
10. Baer, N. S., Majewski, L. J., "Ivory and Related Materials: An Annotated Bibliography, Section A: Conservation and Scientific Examination," *Art Archaeol. Tech. Abstr.* (1970) **8**, 229-275.
11. Weiner, J. S., Oakley, K. P., Le Gros Clark, W. E., "The Solution of the Piltdown Problem," *Geology* (1953) **2**, 139-146.

12. Oakley, K. P., Weiner, J. S., "Piltdown Man," *Amer. Scientist* (1955) **43,** 573-583.
13. Middleton, J., "On Fluorine in Bones, its Source, and its Application to the Determination of the Geological Age of Fossil Bones," *Proc. Geol. Soc. London* (1844) **4,** Pt. 2 (99), 431-433.
14. Carnot, A., "Researches on Fluorine in Modern Bone and in Fossil Bone," *Compt. Rend. Acad. Sci.* (1892) **114,** 1189-1192.
15. De Vries, H., Oakley, K. P., "Radiocarbon Dating of the Piltdown Skull and Jaw," *Nature* (1959) **184,** 224-226.
16. Ho, T. Y., Marcus, L. F., Berger, R., "Radiocarbon Dating of Petroleum-Impregnated Bone from Tar Pits at Rancho La Brea, California," *Science* (1969) **164,** 1051-1052.
17. Morrison, R. T., Boyd, R. N., "Organic Chemistry," pp. 242-243, Allyn and Bacon, Boston, 1959.
18. Schwarzkopf, Otto, private communication.
19. Jaffe, E. B., Sherwood, A. M., "Physical and Chemical Comparison of Modern and Fossil Tooth and Bone Material," *U.S. Geol. Surv.* (1951) **T.E.M.-149.**
20. Plenderleith, H. J., Werner, A. E. A., "The Conservation of Antiquities and Works of Art," 2nd ed., Chap. 6, Oxford University Press, London, 1971.
21. Crawford, Vaughn E., private communication.
22. Baer, N. S., Indictor, N., unpublished data.

RECEIVED August 20, 1973.

INDEX

INDEX

A

Abbasid coinage 138
Absorbing particles, general
  scattering theory for 222
Absorption, atomic 174
Absorption, consumptive 220
Acem Huyuk 236, 238
Activation analysis,
  neutron . . 11, 15, 22, 48, 81, 100, 124
  of ancient pottery, biblical
    studies through 48
  of gold in silver 124
Additives in ink, colored 220, 221
Aegean pottery 68, 76
Al-Rayy 132
Amber, structure of 232
American silver, Hispanic– 139
American silver in the European
  economy, role of 139
Analyses
  for copper 17
  elemental 11, 243
  fluoride 243
  for gold 19
  for lead 2, 20
Analysis
  of ancient pottery, biblical studies
    through activation 48
  of archaeological bronzes 148
  for coins, streak method of 127
  elemental 11, 243
  of glasses 182
  lead isotopic 2, 20
  methods for coins 126
  neutron activation 11, 15, 22, 48,
    81, 100, 124
  of Sasanian silver 11
  of variance approach 175
Analytical procedures,
  nondestructive 125
Analyzing a painting 199
Ancient
  ceramics 55
  Greek pottery 204
  metal 148
  Near Eastern archaeological
    artifacts 236
  pottery, biblical studies through
    activation analysis of 48
  pottery, MES used to study ... 201
  pottery and modern clay,
    correlation between 88
  walrus tusk 239
  world, mining regions in the .. 1
Animal fat, hydrolyzed 231
Antimony 27
Archaeological
  bronzes, comparative analysis of 148
  ceramic study 34
  clays and sherds, testing 37
  sites and clay sources 69
  test problems 41
Archaeology, NMR spectrometry in 226
Argentiferous lead ores 23
Argentite 131
Armenia 132
Arsenic 27
Art objects, metallic 124
Artist's pigments 186
Assyrian ivory 240
Atomic absorption 174
  flameless 130
Attic ware 66
Authenticity for ivories 237

B

Babylon 57
Back-scatter detector for MES ... 190
Back-scatter MES, uses of 204
Barcelona 140
Becerra clay formation 97
Beefbone 239
Biblical studies through activation
  analysis of ancient pottery ... 48
Bismuth 142
Black cores, pottery with dark
  gray to 42
Black glaze ware 66, 68
Black pigment found on a Shang
  oracle bone 208
Bluish ink, modern 211
Boar tooth 239
British Museum 136
Bromine 27
Bronze(s)
  comparative analysis of
    archaeological 148
  horse in Metropolitan Museum
    of Art 1
  low-tin 175
  Luristan 182
Byzantine miliaresia 138

C

Carbon
  combustion 216
  manufacture 217

Carbon *(Continued)*
  particles 216
  pigment 209
Ceramic(s)
  ancient 55
  archaeological 34
  Greek 88
  Hesi 65
  Near Eastern 57
  Oxacan 93
Cerussite ($PbCO_3$) 23
Char 216, 218
Chartres Cathedral 100
  stained glass from 101
Chateau of Rouen, glass specimens
    from 108
Chemical shift 188, 227
Chemistry, wet 175
Chinese ink
  in Far Eastern paintings 207
  general features of 216
  in Korea 210
  origins of 208
  technical studies on 210
  tint of 220
Chinese literature on ink 208
Chromium 27, 55, 88, 117
  in clay 55
Church windows at
    Monkwearmouth 104
Clay(s) 199
  bodies, color changes in 35
  chromium in 55
  cobalt in 55
  correlation between the ancient
    pottery and modern 88
  Egyptian 52
    Nile 72
  europium in 55
  firing tests of 36
  fluxing action on 42
  formation, becerra 97
  Gaza 52
  Hebron 53
  iron in 55
  iron:scandium correlation in ... 55
  kaolin 56
  limestone-associated 51
  red 52
  scandium in 55
  and sherds, testing archaeological 37
  sources, archaeological sites and 69
  studies, Egyptian 45
  types, Palestinian 53
  yellow 52
Cleveland Museum of Art 12
Cobalt 27, 55, 117
  in clay 55
Coinage
  Abbasid 138
  Khusrau II 134
  Umayyad 131
  Sasanian 133
  of Zapata, silver 142
Coins 7
  analysis methods for 126
  Elizabethan 141
  gold impurity in silver 124
  Sasanian 134
  streak method of analysis for ... 127
Collagen 239
Color changes in clay bodies 35
Colored additives 220
Colors in ancient pottery 34
Combustion carbon 216
Comparative analysis of
    archaeological bronzes 148
Computer program 87, 88, 107
Constants, coupling 227
Consumptive absorption 220, 223
Copper 11, 170
  analyses 17
  and gold in Sasanian silver 27
  surface depletion of 128
Corning Museum of Glass 228
Correlation between ancient pottery
    and modern clay 88
Correlation coefficient, Spearman's
    rank 171
Coupling constants 227
Coyotepec 93
  Rosa of 83
Criteria for judging the authenticity
    of Sasanian silver 138
Cupellation 23
p-Cymene 234
Cypriote sherds 52, 76

D

Damascus 132
Data processing 81
Dead Sea salt 54
Dispersion, light 221
  through scattering 220
Drilling samples 237
  vs. streak samples 128
Drillings, micro- 24

E

Eastern paintings, Chinese ink
    in Far 207
Egyptian clay studies 45, 52
Egyptian Nile clay 72
Electron micrographs, scanning .. 210
  of inks 216
Electron microprobe 11, 15
Electron microscope, scanning 43
Elemental analyses 11, 243
Elephant ivory 239
Elizabethan coins 141
European economy, role of
    American silver in the 139
Europium in clay 55

F

Fat, hydrolyzed animal 231

Ferruginous slip 42
Figurines and pottery from the Valley of Mexico, terra cotta 81
Figurines, Tlatilco 97
Firing temperatures 41
Firing tests of clays 36
Flameless atomic absorption 130
Fluorescence, x-ray 175
Fluoride analyses 243
Fluoride exchange 237
Fluxing action on clay 42
Fossil resins 232
Freer Gallery of Art 12, 211
Furnace, thermal gradient 35

G

Galena (PbS) 23
 ores 131
Gamma-rays, scattered 188
Gaza clay 52
General scattering theory for absorbing particles 222
Genoa 140
Geometry, transmission and scattering 189
Glass
 analysis of 182
 from Chartres Cathedral, stained 101
 Corning Museum of 228
 grisaille 101, 109
 -making during the Middle Ages 106
 Medieval 100
 potassium oxide in 111
 sodium oxide in 106
Goethite 194
Gold 11, 30, 155
 impurity in silver coins 124, 135
 in silver 23
  neutron activation analysis of 124
  Sasanian 27
Grain silos at Hesi, sunken 50
Gray to black cores, pottery with dark 42
Greece 1
Greek ceramics 88
Greek potters 43
Greek pottery, ancient 204
"Green earth" 195
Grisaille glass 101, 109

H

Hasanlu 236, 239
Hebron clay 53
Hematite 194
Hermitage Museum 136
Hesi 57
 ceramics 65
 sunken grain silos at 50
Hippopotamus tooth 239
Hispanic–American silver 139
Hydrolyzed animal fat 231
Hydrolyzed vegetable oil 228
Hyperfine magnetic splitting 189

I

Impurity concentrations in silver objects 23
India (or Indian) ink 208
Ink
 additives in 221
 Chinese 216
  in Far Eastern paintings 207
  in Korea 210
  literature on 208
  origins of 208
  technical studies on 210
  tint of 220
 India (or Indian) 208
 Japanese 208
 method for making 209
 modern 218
  bluish 211
  lampblack 211
 scanning electron micrographs of 216
Integral 227
Interlaboratory studies 179
Iran 52, 133
Iridium 23, 27, 31
Iron 27, 88, 117, 175
 in clay 55
 and its compounds 186
 oxide pigments 194
 :scandium correlation in clay .. 55
Irradiation, neutron 87
 Howitzer 127
Isotopic analysis of Laurion lead ores 1
Ivories, authenticity for 237
Ivory, Assyrian 240

J

Japanese ink 208
Jarmo 37
Jayy 132
Jerusalem 70

K

Kaolin clays 56
Khorsabad 237, 240
Khusrau II coinage 134
Korea, Chinese ink in 210

L

Lambityeco 93
Lampblack 208, 210, 219, 220
 ink, modern 211
Laurion lead ores, isotopic analysis of 1
Laurion mines 6
Lead 1, 11, 142, 175
 analyses for 20
 ores 3
  argentiferous 23
  isotopic analysis of Laurion .. 1, 2
  silver from 130
Light dispersion 221
 through scattering 220

Limestone-associated clays 51
Low-tin bronze 175
Luristan bronzes 182

M

Magnetic resonance spectrometry in archaeology, nuclear 226
Magnetite 195
Mahalanobis distances 87
Major and trace elements in Sasanian silver 22
Mammoth tusk 239
Manganese 27
Map of the Near East 69
Mass spectrometry 1, 174
Mastadon 239
Matching technique, relative fit .. 55
Medieval glass 100
Mercury 27
Merv 132
Mesoamerica, preclassic sites in .. 86
Mesopotamian sherds 52
Metal, ancient 148
Metallic art objects 124
  Sasanian 135
Metallurgical structure 12
Method for making ink 209
Methyl palmitate 233
Mexico, Precolumbian Valley of .. 81
  terra cotta figurines and pottery from 81
Microdrillings 24
Microsampling techniques 24
Microscope, scanning electron 43
Middle Ages, glassmaking during the 106
Milan 140
Mining regions in the ancient world 1
Mixteca Puebla polychrome 93
Modern clay, correlation between ancient pottery and 88
Modern ink 218
  bluish 211
  lampblack 211
Modern pottery 52
Moessbauer effect spectroscopy (MES) 186, 204
  back-scatter detector for 190
  nondestructive 186
    to study ancient pottery 201
Monkwearmouth, church windows at 104
Morotoria 71
Mudbrick samples 51
Muds, Nile 52, 55
Multivariate statistical procedures 87
Museum
  British 136
  Cleveland 12
  The Cloisters 109
  Hermitage 136
  Newark 228
  Tehran Archaeological 136

N

Near East, map of 69
Near Eastern archaeological ivory artifacts, ancient 236
Near Eastern ceramics 57
Neutron activation analysis 11, 15, 48, 81, 100
  of gold in silver 124
Neutron irradiation 87
  Howitzer 127
Newark Museum 228
Nickel 27, 88
Nile clay 72
Nile muds 43, 52, 55
Nimrud 236
Nondestructive analytical procedures 125
Nondestructive MES 186
Nuclear magnetic resonance spectrometry in archaeology 226

O

Obsidian 88
Olmec trade 86
Oil, hydrolyzed vegetable 228
Oleic acid 229
Ores, lead 3
  isotopic analysis of 1
Origins of Chinese ink 208
Oxacan ceramics 83, 93
Oxide concentrations in pottery .. 90

P

Painting, analyzing a 199
Paintings, Chinese ink in Far Eastern 207
Palestine 50
Palestinian clay types 53
Particle size distribution 211, 219
  variability of 223
Particles, general scattering theory for absorbing 222
Persian Empire 50
Persian period 49
Pigment(s)
  artist's 186
  carbon 209
  found on a Shang oracle bone, black 208
  iron oxide 194
Polarography 175
Potassium 27
  oxide in glass 111
Potosí silver 140
POTPLOT computer program 55, 87
Potsherds, pre-Han 208
POTSTAT computer program 88
Potters, Greek 43
Pottery
  Aegean 68
  ancient Greek 204
  colors in ancient 34

Pottery *(Continued)*
with dark gray to black cores .. 42
MES to study ancient 201
modern 52
from the Valley of Mexico, terra cotta figurines and 81
Preclassic sites in Mesoamerica ... 86
Precolumbian Valley of Mexico .. 81
Pre-Han potsherds 208
"Price Revolution" of the sixteenth century 139
Princeton University Art Museum 121
Program, computer 55, 87, 88, 107
Prussian blue 199
Puebla polychrome, Mixteca 93

Q

Quadrupole splitting 189

R

"*r*" test 171
Rayy 131
Red clay 52
Relative fit matching technique ... 55
Reproducibility of measurements .. 215
Resins, fossil 232
Restoration, sampling during 109
Role of American silver in the European economy 139
Rosa of Coyotepec 83
Rouen, glass from the Chateau of 108
Rubbings, surface 24

S

Salt, Dead Sea 54
Sampling 12, 27
micro-, techniques of 24
during restoration 109
Sand temper 54
Sasanian
art objects, silver 100
criteria for judging the authenticity of 138
coinage 133
silver
analysis of 11
copper and gold in 27
major and trace elements in .. 22
style silver plates 137
Scandium 27
:iron correlation in clay 55
Scanning electron micrographs of inks 210, 216
Scanning electron microscope 43
Scattered gamma-rays 188
Scattering geometry, transmission and 189
Scattering, light dispersion through 220
Scattering theory for absorbing particles, general 222
Selenium 27
Shang oracle bone, black pigment found on a 208
Sherds 51
Aegean 76
Cypriote 76
Mesopotamian 52
Oaxacan-type 83
Teotihuacan 82
testing archaeological clays and 37
Silos at Hesi, sunken grain 50
Silver
analysis of Sasanian 11
coinage of Zapata 142
coins, gold impurity in 124
in the European economy, role of American 139
fineness 134
gold in 23
Hispanic–American 139
from lead ores 130
objects, impurity concentrations in 23
plates, Sasanian-style 137
Potosi 140
Sasanian 136
copper and gold in 27
criteria for judging the authenticity of 138
major and trace elements in .. 22
silver ratios in 130
Sites and clay sources, archaeological 69
Sixteenth century, "Price Revolution" of the 139
Slip, ferruginous 42
Sodium 27
oxide in glass 106
Soot 208, 210, 220
Spearman's rank correlation coefficient 171
Spectrometry in archaeology, NMR 226
Spectrometry, mass 174
Spectroscopy without sampling, Moessbauer effect 186, 204
Sperm whale tooth 200
Spin-spin splitting 227
Stained glass from Chartres Cathedral 101
Statistical procedures, multivariate 87
Streak method of analysis for coins 127
Streak samples, drilling samples *vs.* 128
Structure of amber 232
Structure, metallurgical 12
Sumi 208
Sunken grain silos at Hesi 50
Surface depletion of copper 128
Supermagnetism 199
Surface rubbings 24
Syria 50, 52, 228

T

Taxonomy 81
Tlatilco figurines 81, 97
Tehran Archaeological Museum .. 136

Tell Beit Mirsim 63
Tell Gezer 49
Tell Halaf 42
Tell-el-Hesi 49, 51, 65
Temperatures, firing 41
Teotihuacan sherds 82
Terra cotta 200
  figurines and pottery from the Valley of Mexico 81
  statuary 186
Test problems, archaeological 41
Testing archaeological clays and sherds 37
Thermal gradient furnace 25
Thermal neutron activation analysis 11, 15, 22
Thermocouple probe measurements 42
Thorium 117
Tin 27
Tint of Chinese ink 220
Trace elements in Sasanian silver, major and 22
Transmission and scattering geometry 189
Triacetin 230
Turkey 52

U

Umayyad coinage 131

V

Valley of Mexico, Precolumbian .. 81
  terra cotta figurines and pottery from 81
Variability of particle size distribution 223
Vegetable oil, hydrolyzed 228

W

Wei dynasty 208
Wet chemistry 175
Whale bone 239
Wood ash 106

X

X-ray fluorescence 11, 15, 175

Y

Yellow clay 52
Youden plot 171

Z

Zapata, silver coinage of 142
Zinc 27, 175

The text of this book is set in 10 point Caledonia with two points of leading. The chapter numerals are set in 30 point Garamond; the chapter titles are set in 18 point Garamond Bold.

The book is printed offset on Glatfelter Old Forge Enamel, 80-pound. The cover is Joanna Book Binding blue linen.

Editing and production by Mary Westerfeld.

The book was composed by the Mills-Frizell-Evans Co., printed by The McFarland Co., Harrisburg, Pa., and bound by The Maple Press Co., York, Pa.